ATF Federal Explosives Law and Regulations

Plus

ATF Firearms Curios or Relics List

ATF Federal Explosives Law and Regulations

Table of Contents

Organized Crime Control Act of 1970, Title XI

Regulation of Explosives

Public Law 91-452, Approved October 15, 1970 (as Amended)

[**NOTE:** Any reference to the Internal Revenue Code of 1954 refers to the Internal Revenue Code of 1986 (§2, Public Law 99-514, 100 Stat. 2085, October 22, 1986.)]

Purpose

SEC. 1101. The Congress hereby declares that the purpose of this title is to protect interstate and foreign commerce against interference and interruption by reducing the hazard to persons and property arising from misuse and unsafe or insecure storage of explosive materials. It is not the purpose of this title to place any undue or unnecessary Federal restrictions or burdens on law-abiding citizens with respect to the acquisition, possession, storage, or use of explosive materials for industrial, mining, agricultural, or other lawful purposes, or to provide for the imposition by Federal regulations of any procedures or requirements other than those reasonably necessary to implement and effectuate the provisions of this title.

Sec. 1102. Title 18, United States Code, is amended by adding after chapter 39 the following chapter:

Chapter 40. Importation, Manufacture, Distribution and Storage of Explosive Materials

Editor's Note: The sections of law set out herein were added by Public Law 91-452, Title XI, §1102(a), Oct. 15, 1970, 84 Stat. 952-959, and remain un-changed unless otherwise footnoted.

Sec.

841. Definitions
842. Unlawful acts
843. Licenses and user permits
844. Penalties
845. Exceptions; relief from disabilities
846. Additional powers of the Attorney General
847. Rules and regulations
848. Effect on State law

§ 841. Definitions

As used in this chapter

(a) "Person" means any individual, corporation, company, association, firm, partnership, society, or joint stock company.

(b) "Interstate or foreign commerce" means commerce between any place in a State and any place outside of that State, or within any possession of the United States (not including the Canal Zone) or the District of Columbia, and commerce between places within the same State but through any place outside of that State. "State" includes the District of Columbia, the Commonwealth of Puerto Rico, and the possessions of the United States (not including the Canal Zone).

(c) "Explosive materials" means explosives, blasting agents, and detonators.

(d) Except for the purposes of subsections (d), (e), (f), (g), (h), (i), and (j) of section 844 of this title, **"explosives"** means any chemical compound mixture, or device, the primary or common purpose of which is to function by explosion; the term includes, but is not limited to, dynamite and other high explosives, black powder, pellet powder, initiating explosives, detonators, safety fuses, squibs, detonating cord, igniter cord, and igniters. The Attorney General shall publish and revise at least annually in the Federal Register a list of these and any additional explosives which he determines to be within the coverage of this chapter. For the purposes of subsections (d), (e), (f), (g), (h), and (i) of section 844 of this title, the term "explosive" is defined in subsection (j) of such section 844.

(e) "Blasting agent" means any material or mixture, consisting of fuel and oxidizer, intended for blasting, not otherwise defined as an explosive: Provided, That the finished product, as mixed for use or shipment, cannot be detonated by means of a numbered 8 test blasting cap when unconfined.

(f) "Detonator" means any device containing a detonating charge that is used for initiating detonation in an explosive; the term includes, but is not limited to, electric blasting caps of instantaneous and delay types, blasting caps for use with safety fuses and detonating-cord delay connectors.

(g) "Importer" means any person engaged in the business of importing or bringing explosive materials into the United States for purposes of sale or distribution.

(h) "Manufacturer" means any person engaged in the business of manufacturing explosive materials for purposes of sale or distribution or for his own use.

(i) "Dealer" means any person engaged in the business of distributing explosive materials at wholesale or retail.

(j) "Permittee" means any user of explosives for a lawful purpose, who has obtained either a user permit or a limited permit under the provisions of this chapter.

(k) "Attorney General" means the Attorney General of the United States.

(l) "Crime punishable by imprisonment for a term exceeding one year" shall not mean (1) any Federal or State offenses pertaining to antitrust violations, unfair trade practices, restraints of trade, or other similar offenses relating to the regulation of business practices as the Attorney General may be regulation designate, or (2) any State offense (other than one involving a firearm or explosive) classified by the laws of the State as a misdemeanor and punishable by a term of imprisonment of two years or less.

(m) "Licensee" means any importer, manufacturer, or dealer licensed under the provisions of this chapter.

(n) "Distribute" means sell, issue, give, transfer, or otherwise dispose of.

(o) "Convention on the Marking of Plastic Explosives" means the Convention on the Marking of Plastic Explosives for the Purpose of Detection, Done at Montreal on 1 March 1991.

(p) "Detection agent" means any one of the substances specified in this subsection when introduced into a plastic explosive or formulated in such explosive as a part of the manufacturing process in such a manner as to achieve homogeneous distribution in the finished explosive, including—

(1) Ethylene glycol dinitrate (EGDN), $C_2H_4(NO_3)_2$, molecular weight 152, when the minimum concentration in the finished explosive is 0.2 percent by mass;

(2) 2,3-Dimethyl-2,3-dinitrobutane (DMNB), $C_6H_{12}(NO_2)_2$, molecular weight 176, when the minimum concentration in the finished explosive is 0.1 percent by mass;

(3) Para-Mononitrotoluene (p-MNT), $C_7H_7NO_2$, molecular weight 137, when the minimum concentration in the finished explosive is 0.5 percent by mass;

(4) Ortho-Mononitrotoluene (o-MNT), $C_7H_7NO_2$, molecular weight 137, when the minimum concentration in the finished explosive is 0.5 percent by mass; and

(5) any other substance in the concentration specified by the Attorney General, after consultation with the Secretary of State and the Secretary of Defense, that has been added to the table in part 2 of the Technical Annex to the Convention on the Marking of Plastic Explosives.

(q) "Plastic explosive" means an explosive material in flexible or elastic sheet form formulated with one or more high explosives which in their pure form has a vapor pressure less than 10^{-4} Pa at a temperature of 25 degrees C., is formulated with a binder material, and is as a mixture malleable or flexible at normal room temperature.

(r) "Alien" means any person who is not a citizen or national of the United States.

(s) "Responsible person" means an individual who has the power to direct the management and policies of the applicant pertaining to explosive materials.

(Added Pub. L. 91-452, title XI, Sec. 1102(a), Oct. 15, 1970, 84 Stat. 952; amended Pub. L. 104-132, title VI, Sec. 602, Apr. 24, 1996, 110 Stat. 1288; Pub. L. 107-296, title XI, Sec. 1112(e)(1), (3), 1122(a), Nov. 25, 2002, 116 Stat. 2276, 2280.)

§ 842. Unlawful acts

(a) It shall be unlawful for any person—

(1) to engage in the business of importing, manufacturing, or dealing in explosive materials without a license issued under this chapter;

(2) knowingly to withhold information or to make any false or fictitious oral or written statement or to furnish or exhibit any false, fictitious, or misrepresented identification, intended or likely to deceive for the purpose of obtaining explosive materials, or a license, permit, exemption, or relief from disability under the provisions of this chapter;

(3) other than a licensee or permittee knowingly–

(A) to transport, ship, cause to be transported, or receive any explosive materials; or

(B) to distribute explosive materials to any person other than a licensee or permittee; or

(4) who is a holder of a limited permit—

(A) to transport, ship, cause to be transported, or receive in interstate or foreign commerce any explosive materials; or

(B) to receive explosive materials from a licensee or permittee, whose premises are located outside the State of residence of the limited permit holder, or on more than 6 separate occasions, during the period of the permit, to receive explosive materials from 1 or more licensees or permittees whose premises are located within the State of residence of the limited permit holder.

(b) It shall be unlawful for any licensee or permittee to knowingly distribute any explosive materials to any person other than—

(1) a licensee;

(2) a holder of a user permit; or

(3) a holder of a limited permit who is a resident of the State where distribution is made and in which the premises of the transferor are located.

(c) It shall be unlawful for any licensee to distribute explosive materials to any person who the licensee has reason to believe intends to transport such explosive materials into a State where the purchase, possession, or use of explosive materials is prohibited or which does not permit its residents to transport or ship explosive materials into it or to receive explosive materials in it.

(d) It shall be unlawful for any person knowingly to distribute explosive materials to any individual who:

(1) is under twenty-one years of age;

(2) has been convicted in any court of a crime punishable by imprisonment for a term exceeding one year;

(3) is under indictment for a crime punishable by imprisonment for a term exceeding one year;

(4) is a fugitive from justice;

(5) is an unlawful user of or addicted to any controlled substance (as defined in section 102 of the Controlled Substances Act);

(6) has been adjudicated a mental defective or who has been committed to a mental institution;

(7) is an alien, other than an alien who—

(A) is lawfully admitted for permanent residence (as defined in section 101 (a)(20) of the Immigration and Nationality Act);

(B) is in lawful nonimmigrant status, is a refugee admitted under section 207 of the Immigration and Nationality Act or is in asylum status under section 208 of the Immigration and Nationality Act and—

(i) is a foreign law enforcement officer of a friendly foreign government, as determined by the Secretary in consultation with the Secretary of State, entering the United States on official law enforcement business, and the shipping, transporting, possession, or receipt of explosive materials is in furtherance of this official law enforcement business; or

(ii) is a person having the power to direct or cause the direction of the management and policies of a corporation, partnership, or association licensed pursuant to section 843(a), and the shipping, transporting, possession, or receipt of explosive materials is in furtherance of such power;

(C) is a member of a North Atlantic Treaty Organization (NATO) or other friendly foreign military force, as determined by the Attorney General in consultation with the Secretary of Defense, who is present in the United States under military orders for training or other military purpose authorized by the United States and the shipping, transporting, possession, or receipt of explosive materials is in furtherance of the authorized military purpose; or

(D) is lawfully present in the United States in cooperation with the Director of Central Intelligence, and the shipment, transportation, receipt, or possession of the explosive materials is in furtherance of such cooperation;

(8) has been discharged from the armed forces under dishonorable conditions;

(9) having been a citizen of the United States, has renounced the citizenship of that person.

(e) It shall be unlawful for any licensee knowingly to distribute any explosive materials to any person in any State where the purchase, possession, or use by such person of such explosive materials would be in violation of any State law or any published ordinance applicable at the place of distribution.

(f) It shall be unlawful for any licensee or permittee willfully to manufacture, import, purchase, distribute, or receive explosive materials without making such records as the Attorney General may by regulation require, including, but not limited to, a statement of intended use, the name, date, place of birth, social security number or taxpayer identification number, and place of residence of any natural person to whom explosive materials are distributed. If explosive materials are distributed to a corporation or other business entity, such records shall include the identity and principal and local places of business and the name, date, place of birth, and place of residence of the natural person acting as agent of the corporation or other business entity in arranging the distribution.

(g) It shall be unlawful for any licensee or permittee knowingly to make any false entry in any record which he is required to keep pursuant to this section or regulations promulgated under section 847 of this title.

(h) It shall be unlawful for any person to receive, possess, transport, ship, conceal, store, barter, sell, dispose of, or pledge or accept as security for a loan, any stolen explosive materials which are moving as, which are part of, which constitute, or which have been shipped or transported in, interstate or foreign commerce, either before or after such materials were stolen, knowing or having reasonable cause to believe that the explosive materials were stolen.

(i) It shall be unlawful for any person—

(1) who is under indictment for, or who has been convicted in any court of, a crime punishable by imprisonment for a term exceeding one year;

(2) who is a fugitive from justice;

(3) who is an unlawful user of or addicted to any controlled substance (as defined in section 102 of the Controlled Substances Act);

(4) who has been adjudicated as a mental defective or who has been committed to a mental institution;

(5) who is an alien, other than an alien who—

(A) is lawfully admitted for permanent residence (as that term is defined in section 101(a)(20) of the Immigration and Nationality Act);

(B) is in lawful nonimmigrant status, is a refugee admitted under section 207 of the Immigration and Nationality Act or is in asylum status under section 208 of the Immigration and Nationality Act and--

(i) is a foreign law enforcement officer of a friendly foreign government, as determined by the Secretary in consultation with the Secretary of State, entering the United States on official law enforcement business, and the shipping, transporting, possession, or receipt of explosive materials is in furtherance of this official law enforcement business; or

(ii) is a person having the power to direct or cause the direction of the management and policies of a corporation, partnership, or association licensed pursuant to section 843(a), and the shipping, transporting, possession, or receipt of explosive materials is in furtherance of such power;

(C) is a member of a North Atlantic Treaty Organization (NATO) or other friendly foreign military force, as determined by the Attorney General in consultation with the Secretary of Defense, who is present in the United States under military orders for training or other military purpose authorized by the United States and the shipping, transporting, possession, or receipt of explosive materials is in furtherance of the authorized military purpose; or

(D) is lawfully present in the United States in cooperation with the Director of Central Intelligence, and the shipment, transportation, receipt, or possession of the explosive materials is in furtherance of such cooperation;

(6) who has been discharged from the armed forces under dishonorable conditions;

(7) who, having been a citizen of the United States, has renounced the citizenship of that person; to ship or transport any explosive in or affecting interstate or foreign commerce or to receive or possess any explosive which has been shipped or transported in or affecting interstate or foreign commerce.

(j) It shall be unlawful for any person to store any explosive material in a manner not in conformity with regulations promulgated by the Attorney General. In promulgating such regulations, the Attorney General shall take into consideration the class, type, and quantity of explosive materials to be stored, as well as the standards of safety and security recognized in the explosives industry.

(k) It shall be unlawful for any person who has knowledge of the theft or loss of any explosive materials from his stock, to fail to report such theft or loss within twenty-four hours of discovery thereof, to the Attorney General and to appropriate local authorities.

(l) It shall be unlawful for any person to manufacture any plastic explosive that does not contain a detection agent.

(m) **(1)** It shall be unlawful for any person to import or bring into the United States, or export from the United States, any plastic explosive that does not contain a detection agent.

(2) This subsection does not apply to the importation or bringing into the United States, or the exportation from the United States, of any plastic explosive that was imported or brought into, or manufactured in the United States prior to the date of enactment of this subsection [enacted April 24, 1996] by or on behalf of any agency of the United States performing military or police functions (including any military reserve component) or by or on behalf of the National Guard of any State, not later than 15 years after the date of entry into force of the Convention on the Marking of Plastic Explosives, with respect to the United States.

(n) **(1)** It shall be unlawful for any person to ship, transport, transfer, receive, or possess any plastic explosive that does not contain a detection agent.

(2) This subsection does not apply to—

(A) the shipment, transportation, transfer, receipt, or possession of any plastic explosive that was imported or brought into, or manufactured in the United States prior to the date of enactment of this subsection [enacted April 24, 1996] by any person during the period beginning on that date and ending 3 years after that date of enactment; or

(B) the shipment, transportation, transfer, receipt, or possession of any plastic explosive that was imported or brought into, or manufactured in the United States prior to the date of enactment of this subsection [enacted April 24, 1996] by or on behalf of any agency of the United States performing a military or police function (including any military reserve component) or by or on behalf of the National Guard of any State, not later than 15 years after the date of entry into force of the Convention on the Marking of Plastic Explosives, with respect to the United States.

(o) It shall be unlawful for any person, other than an agency of the United States (including any military reserve component) or the National Guard of any State, possessing any plastic explosive on the date of enactment of this subsection [enacted April 24, 1996], to fail to report to the Attorney General within 120 days after such date of enactment the quantity of such explosives possessed, the manufacturer or importer, any marks of identification on such explosives, and such other information as the Attorney General may prescribe by regulation.

(p) Distribution of information relating to explosives, destructive devices, and weapons of mass destruction.

(1) Definitions. In this subsection—

(A) the term "destructive device" has the same meaning as in section 921(a)(4);

(B) the term "explosive" has the same meaning as in section 844(j); and

(C) the term "weapon of mass destruction" has the same meaning as in section 2332a(c)(2).

(2) Prohibition. It shall be unlawful for any person—

(A) to teach or demonstrate the making or use of an explosive, a destructive device, or a weapon of mass destruction, or to distribute by any means information pertaining to, in whole or in part, the manufacture or use of an explosive, destructive device, or weapon of mass destruction, with the intent that the teaching, demonstration, or information be used for, or in furtherance of, an activity that constitutes a Federal crime of violence; or

(B) to teach or demonstrate to any person the making or use of an explosive, a destructive device, or a weapon of mass destruction, or to distribute to any person, by any means, information pertaining to, in whole or in part, the manufacture or use of an explosive, destructive device, or weapon of mass destruction, knowing that such person intends to use the teaching, demonstration, or information for, or in furtherance of, an activity that constitutes a Federal crime of violence.

(Added Pub. L. 91-452, title XI, Sec. 1102(a), Oct. 15, 1970, 84 Stat. 953; amended Pub. L. 100-690, title VI, Sec. 6474(c), (d), Nov. 18, 1988, 102 Stat. 4380; Pub. L. 101-647, title XXXV, Sec. 3521, Nov. 29, 1990, 104 Stat. 4923; Pub. L. 103-322, title XI, Sec. 110508, 10516, Sept. 13, 1994, 108 Stat. 2018, 2020; Pub. L. 104-132, title VI, Sec. 603, title VII, Sec. 707, Apr. 24, 1996, 110 Stat. 1289, 1296; Pub. L. 106-54, Sec. 2(a), Aug. 17, 1999, 113 Stat. 398; Pub. L. 107-296, title XI, Sec. 1112(e)(3), 1122(b), 1123, Nov. 25, 2002, 116 Stat. 2276, 2280, 2283.)

§ 843. Licenses and user permits

(a) An application for a user permit or limited permit or a license to import, manufacture, or deal in explosive materials shall be in such form and contain such information as the Attorney General shall by regulation prescribe, including the names of and appropriate identifying information regarding all employees who will be authorized by the applicant to possess explosive materials, as well as fingerprints and a photograph of each responsible person. Each applicant for a license or permit shall pay a fee to be charged as set by the Attorney General, said fee not to exceed $ 50 for a limited permit and $ 200 for any other license or permit. Each license or user permit shall be valid for not longer than 3 years from the date of issuance and each limited permit shall be valid for not longer than 1 year from the date of issuance. Each license or permit shall be renewable upon the same conditions and subject to the same restrictions as the original license or permit, and upon payment of a renewal fee not to exceed one-half of the original fee.

(b) Upon the filing of a proper application and payment of the prescribed fee, and subject to the provisions of this chapter and other applicable laws, the Attorney General shall issue to such applicant the appropriate license or permit if--

(1) the applicant (or, if the applicant is a corporation, partnership, or association, each responsible person with respect to the applicant) is not a person described in section 842(i);

(2) the applicant has not willfully violated any of the provisions of this chapter or regulations issued hereunder;

(3) the applicant has in a State premises from which he conducts or intends to conduct business;

(4) **(A)** the Secretary verifies by inspection or, if the application is for an original limited permit or the first or second renewal of such a permit, by such other means as the Secretary determines appropriate, that the applicant has a place of storage for explosive materials which meets such standards of public safety and security against theft as the Attorney General by regulations shall prescribe; and

(B) subparagraph (A) shall not apply to an applicant for the renewal of a limited permit if the Secretary has verified, by inspection within the preceding 3 years, the matters described in subparagraph (A) with respect to the applicant; and

(5) the applicant has demonstrated and certified in writing that he is familiar with all published State laws and local ordinances relating to explosive materials for the location in which he intends to do business;

(6) none of the employees of the applicant who will be authorized by the applicant to possess explosive materials is any person described in section 842(i); and

(7) in the case of a limited permit, the applicant has certified in writing that the applicant will not receive explosive materials on more than 6 separate occasions during the 12-month period for which the limited permit is valid.

(c) The Attorney General shall approve or deny an application within a period of 90 days for licenses and permits beginning on the date such application is received by the Attorney General.

(d) The Attorney General may revoke any license or permit issued under this section if in the opinion of the Attorney General the holder thereof has violated any provision of this chapter or any rule or regulation prescribed by the Attorney General under this chapter, or has become ineligible to acquire explosive materials under section 842(d). The Attorney General's action under this subsection may be reviewed only as provided in subsection (e)(2) of this section.

(e) **(1)** Any person whose application is denied or whose license or permit is revoked shall receive a written notice from the Attorney General stating the specific grounds upon which such denial or revocation is based. Any notice of a revocation of a license or permit shall be given to the holder of such license or permit prior to or concurrently with the effective date of the revocation.

(2) If the Attorney General denies an application for, or revokes a license, or permit, he shall, upon request by the aggrieved party, promptly hold a hearing to review his denial or revocation. In the case of a revocation, the Attorney General may upon a request of the holder stay the effective date of the revocation. A hearing under this section shall be at a location convenient to the aggrieved party. The Attorney General shall give written notice of his decision to the aggrieved party within a reasonable time after the hearing. The aggrieved party may, within sixty days after receipt of the Secretary's written decision, file a petition with the United States court of appeals for the district in which he resides or has his principal place of business for a judicial review of such denial or revocation, pursuant to sections 701-706 of title 5, United States Code.

(f) Licensees and holders of user permits shall make available for inspection at all reasonable times their records kept pursuant to this chapter or the regulations issued hereunder, and licensees and permittees shall submit to the Attorney General such reports and information with respect to such records and the contents thereof as he shall by regulations prescribe. The Attorney General may enter during business hours the premises (including places of storage) of any licensee or holder of a user permit, for the purpose of inspecting or examining (1) any records or documents required to be kept by such licensee or permittee, under the provisions of this chapter or regulations issued hereunder, and (2) any explosive materials kept or stored by such licensee or permittee at such premises. Upon the request of any State or any political subdivision thereof, the Attorney General may make available to such State or any political subdivision thereof, any information which he may obtain by reason of the provisions of this chapter with respect to the identification of persons within such State or political subdivision thereof, who have purchased or received explosive materials, together with a description of such explosive materials. The Secretary may inspect the places of storage for explosive materials of an applicant for a limited permit or, at the time of renewal of such permit, a holder of a limited permit, only as provided in subsection (b)(4).

(g) Licenses and user permits issued under the provisions of subsection (b) of this section shall be kept posted and kept available for inspection on the premises covered by the license and permit.

(h) **(1)** If the Secretary receives, from an employer, the name and other identifying information of a responsible person or an employee who will be authorized by the employer to possess explosive materials in the course of employment with the employer, the Secretary shall determine whether the responsible person or employee is one of the persons described in any paragraph of section 842(i). In making the determination, the Secretary may take into account a letter or document issued under paragraph (2).

(2) **(A)** If the Secretary determines that the responsible person or the employee is not one of the persons described in any paragraph of section 842(i), the Secretary shall notify the employer in writing or electronically of the determination and issue, to the responsible person or employee, a letter of clearance, which confirms the determination.

(B) If the Secretary determines that the responsible person or employee is one of the persons described in any paragraph of section 842(i), the Secretary shall notify the employer in writing or electronically of the determination and issue to the responsible person or the employee, as the case may be, a document that—

(i) confirms the determination;

(ii) explains the grounds for the determination;

(iii) provides information on how the disability may be relieved; and

(iv) explains how the determination may be appealed.

(i) Furnishing of samples.

(1) In general. Licensed manufacturers and licensed importers and persons who manufacture or import explosive materials or ammonium nitrate shall, when required by letter issued by the Secretary, furnish—

(A) samples of such explosive materials or ammonium nitrate;

(B) information on chemical composition of those products; and

(C) any other information that the Secretary determines is relevant to the identification of the explosive materials or to identification of the ammonium nitrate.

(2) Reimbursement. The Secretary shall, by regulation, authorize reimbursement of the fair market value of samples furnished pursuant to this subsection, as well as the reasonable costs of shipment.

(Added Pub. L. 91-452, title XI, Sec. 1102(a), Oct. 15, 1970, 84 Stat. 955; amended Pub. L. 107-296, title XI, Sec. 1112(e)(3), 1122(c)-(h), 1124, Nov. 25, 2002, 116 Stat. 2276, 2281, 2282, 2285.)

§ 844. Penalties

(a) Any person who—

(1) violates any of subsections (a) through (i) or (l) through (o) of section 842 shall be fined under this title, imprisoned for not more than 10 years, or both; and

(2) violates subsection (p)(2) of section 842, shall be fined under this title, imprisoned not more than 20 years, or both.

(b) Any person who violates any other provision of section 842 of this chapter shall be fined under this title or imprisoned not more than one year, or both.

(c) **(1)** Any explosive materials involved or used or intended to be used in any violation of the provisions of this chapter or any other rule or regulation promulgated thereunder or any violation of any criminal law of the United States shall be subject to seizure and forfeiture, and all provisions of the Internal Revenue Code of 1986 relating to the seizure, forfeiture, and disposition of firearms, as defined in section 5845(a) of that Code, shall, so far as applicable, extend to seizures and forfeitures under the provisions of this chapter.

(2) Notwithstanding paragraph (1), in the case of the seizure of any explosive materials for any offense for which the materials would be subject to forfeiture in which it would be impracticable or unsafe to remove the materials to a place of storage or would be unsafe to store them, the seizing officer may destroy the explosive materials forthwith. Any destruction under this paragraph shall be in the presence of at least 1 credible witness. The seizing officer shall make a report of the seizure and take samples as the Attorney General may by regulation prescribe.

(3) Within 60 days after any destruction made pursuant to paragraph (2), the owner of (including any person having an interest in) the property so destroyed may make application to the Attorney General for reimbursement of the value of the property. If the claimant establishes to the satisfaction of the Attorney General that—

(A) the property has not been used or involved in a violation of law; or

(B) any unlawful involvement or use of the property was without the claimant's knowledge, consent, or willful blindness, the Attorney General shall make an allowance to the claimant not exceeding the value of the property destroyed.

(d) Whoever transports or receives, or attempts to transport or receive, in interstate or foreign commerce any explosive with the knowledge or intent that it will be used to kill, injure, or intimidate any individual or unlawfully to damage or destroy any building, vehicle, or other real or personal property, shall be imprisoned for not more than ten years, or fined under this title, or both; and if personal injury results to any person, including any public safety officer performing duties as a direct or proximate result of conduct prohibited by this subsection, shall be imprisoned for not more than twenty years or fined under this title, or both; and if death results to any person, including any public safety officer performing duties as a direct or proximate result of conduct prohibited by this subsection, shall be subject to imprisonment for any term of years, or to the death penalty or to life imprisonment.

(e) Whoever, through the use of the mail, telephone, telegraph, or other instrument of interstate or foreign commerce, or in or affecting interstate or foreign commerce, willfully makes any threat, or maliciously conveys false information knowing the same to be false, concerning an attempt or alleged attempt being made, or to be made, to kill, injure, or intimidate any individual or unlawfully to damage or destroy any building, vehicle, or other real or personal property by means of fire or an explosive shall be imprisoned for not more than 10 years or fined under this title, or both.

(f) **(1)** Whoever maliciously damages or destroys, or attempts to damage or destroy, by means of fire or an explosive, any building, vehicle, or other personal or real property in whole or in part owned or possessed by, or leased to, the United States, or any department or agency thereof, or any institution or organization receiving Federal financial assistance, shall be imprisoned for not less than 5 years and not more than 20 years, fined under this title, or both.

(2) Whoever engages in conduct prohibited by this subsection, and as a result of such conduct, directly or proximately causes personal injury or creates a substantial risk of injury to any person, including any public safety officer performing duties, shall be imprisoned for not less than 7 years and not more than 40 years, fined under this title, or both.

(3) Whoever engages in conduct prohibited by this subsection, and as a result of such conduct directly or proximately causes the death of any person, including any public safety officer performing duties, shall be subject to the death penalty, or imprisoned for not less than 20 years or for life, fined under this title, or both.

(g) **(1)** Except as provided in paragraph (2), whoever possesses an explosive in an airport that is subject to the regulatory authority of the Federal Aviation Administration, or in any building in whole or in part owned, possessed, or used by, or leased to, the United States or any department or agency thereof, except with the written consent of the agency, department, or other person responsible for the management of such building or airport, shall be imprisoned for not more than five years, or fined under this title, or both.

(2) The provisions of this subsection shall not be applicable to—

(A) the possession of ammunition (as that term is defined in regulations issued pursuant to this chapter) in an airport that is subject to the regulatory authority of the Federal Aviation Administration if such ammunition is either in checked baggage or in a closed container; or

(B) the possession of an explosive in an airport if the packaging and transportation of such explosive is exempt from, or subject to and in accordance with, regulations of the Pipeline and Hazardous Materials Safety Administration for the handling of hazardous materials pursuant to chapter 51 of title 49.

(h) Whoever—

(1) uses fire or an explosive to commit any felony which may be prosecuted in a court of the United States, or

(2) carries an explosive during the commission of any felony which may be prosecuted in a court of the United States, including a felony which provides for an enhanced punishment if committed by the use of a deadly or dangerous weapon or device shall, in addition to the punishment provided for such felony, be sentenced to imprisonment for 10 years. In the case of a second or subsequent conviction under this subsection, such person shall be sentenced to imprisonment for 20 years. Notwithstanding any other provision of law, the court shall not place on probation or suspend the sentence of any person convicted of a violation of this subsection, nor shall the term of imprisonment imposed under this subsection run concurrently with any other term of imprisonment including that imposed for the felony in which the explosive was used or carried.

(i) Whoever maliciously damages or destroys, or attempts to damage or destroy, by means of fire or an explosive, any building, vehicle, or other real or personal property used in interstate or foreign commerce or in any activity affecting interstate or foreign commerce shall be imprisoned for not less than 5 years and not more than 20 years, fined under this title, or both; and if personal injury results to any person, including any public safety officer performing duties as a direct or proximate result of conduct prohibited by this subsection, shall be imprisoned for not less than 7 years and not more than 40 years, fined under this title, or both; and if death results to any person, including any public safety officer performing duties as a direct or proximate result of conduct prohibited by this subsection, shall also be subject to imprisonment for any term of years, or to the death penalty or to life imprisonment.

(j) For the purposes of subsections (d), (e), (f), (g), (h), and (i) of this section and section 842(p), the term "explosive" means gunpowders, powders used for blasting, all forms of high explosives, blasting materials, fuses (other than electric circuit breakers), detonators, and other detonating agents, smokeless powders, other explosive or incendiary devices within the meaning of paragraph (5) of section 232 of this title, and any chemical compounds, mechanical mixture, or device that contains any oxidizing and combustible units, or other ingredients, in such proportions, quantities, or packing that ignition by fire, by friction, by concussion, by percussion, or by detonation of the compound, mixture, or device or any part thereof may cause an explosion.

(k) A person who steals any explosives materials which are moving as, or are a part of, or which have moved in, interstate or foreign commerce shall be imprisoned for not more than 10 years, fined under this title, or both.

(l) A person who steals any explosive material from a licensed importer, licensed manufacturer, or licensed dealer, or from any permittee shall be fined under this title, imprisoned not more than 10 years, or both.

(m) A person who conspires to commit an offense under subsection (h) shall be imprisoned for any term of years not exceeding 20, fined under this title, or both.

(n) Except as otherwise provided in this section, a person who conspires to commit any offense defined in this chapter shall be subject to the same penalties (other than the penalty of death) as the penalties prescribed for the offense the commission of which was the object of the conspiracy.

(o) Whoever knowingly transfers any explosive materials, knowing or having reasonable cause to believe that such explosive materials will be used to commit a crime of violence (as defined in section 924(c)(3)) or drug trafficking crime (as defined in section 924(c)(2)) shall be subject to the same penalties as may be imposed under subsection (h) for a first conviction for the use or carrying of an explosive material.

(p) Theft reporting requirement.

(1) In general. A holder of a license or permit who knows that explosive materials have been stolen from that licensee or permittee, shall report the theft to the Attorney General not later than 24 hours after the discovery of the theft.

(2) Penalty. A holder of a license or permit who does not report a theft in accordance with paragraph (1), shall be fined not more than $ 10,000, imprisoned not more than 5 years, or both.

(Added Pub. L. 91-452, title XI, Sec. 1102(a), Oct. 15, 1970, 84 Stat. 956; amended Pub. L. 97-298, Sec. 2, Oct. 12, 1982, 96 Stat. 1319; Pub. L. 98-473, title II, Sec. 1014, Oct. 12, 1984, 98 Stat. 2142; Pub. L. 99-514, Sec. 2, Oct. 22, 1986, 100 Stat. 2095; Pub. L. 100-690, title VI, Sec. 6474(a), (b), Nov. 18, 1988, 102 Stat. 4379; Pub. L. 101-647, title XXXV, Sec. 3522, Nov. 29, 1990, 104 Stat. 4924; Pub. L. 103-272, Sec. 5(e)(7), July 5, 1994, 108 Stat. 1374; Pub. L. 103-322, title VI, Sec. 60003(a)(3), title XI, Sec. 110504(b), 110509, 110515(b), 110518(b), title XXXII, Sec. 320106, 320917(a), title XXXIII, Sec. 330016(1)(H), (K), (L), (N), Sept. 13, 1994, 108 Stat. 1969, 2016, 2018, 2020, 2111, 2129, 2147, 2148; Pub. L. 104-132, title VI, Sec. 604, title VII, Sec. 701, 706, 708(a), (c)(3), 724, Apr. 24, 1996, 110 Stat. 1289, 1291, 1295-1297, 1300; Pub. L. 104-294, title VI, Sec. 603(a), Oct. 11, 1996, 110 Stat. 3503; Pub. L. 106-54, Sec. 2(b), Aug. 17, 1999, 113 Stat. 399; Pub. L. 107-296, title XI, Sec. 1112(e)(3), 1125, 1127, Nov. 25, 2002, 116 Stat. 2276, 2285.)

§ 845. Exceptions; relief from disabilities

(a) Except in the case of subsections (l), (m), (n), or (o) of section 842 and subsections (d), (e), (f), (g), (h), and (i) of section 844 of this title, this chapter shall not apply to:

(1) aspects of the transportation of explosive materials via railroad, water, highway, or air that pertain to safety, including security, and are regulated by the United States Department of Transportation or the Department of Homeland Security;

(2) the use of explosive materials in medicines and medicinal agents in the forms prescribed by the official United States Pharmacopeia, or the National Formulary;

(3) the transportation, shipment, receipt, or importation of explosive materials for delivery to any agency of the United States or to any State or political subdivision thereof;

(4) small arms ammunition and components thereof;

(5) commercially manufactured black powder in quantities not to exceed fifty pounds, percussion caps, safety and pyrotechnic fuses, quills, quick and slow matches, and friction primers, intended to be used solely for sporting, recreational, or cultural purposes in antique firearms as defined in section 921(a)(16) of title 18 of the United States Code, or in antique devices as exempted from the term "destructive device" in section 921(a)(4) of title 18 of the United States Code; and

(6) the manufacture under the regulation of the military department of the United States of explosive materials for, or their distribution to or storage or possession by the military or naval services or other agencies of the United States; or to arsenals, navy yards, depots, or other establishments owned by, or operated by or on behalf of, the United States.

(b) **(1)** A person who is prohibited from shipping, transporting, receiving, or possessing any explosive under section 842(i) may apply to the Secretary for relief from such prohibition.

(2) The Secretary may grant the relief requested under paragraph (1) if the Secretary determines that the circumstances regarding the applicability of section 842(i), and the applicant's record and reputation, are such that the applicant will not be likely to act in a manner dangerous to public safety and that the granting of such relief is not contrary to the public interest.

(3) A licensee or permittee who applies for relief, under this subsection, from the disabilities incurred under this chapter as a result of an indictment for or conviction of a crime punishable by imprisonment for a term exceeding 1 year shall not be barred by such disability from further operations under the license or permit pending final action on an application for relief filed pursuant to this section.

(c) It is an affirmative defense against any proceeding involving subsections (l) through (o) of section 842 if the proponent proves by a preponderance of the evidence that the plastic explosive—

(1) consisted of a small amount of plastic explosive intended for and utilized solely in lawful—

(A) research, development, or testing of new or modified explosive materials;

(B) training in explosives detection or development or testing of explosives detection equipment; or

(C) forensic science purposes; or

(2) was plastic explosive that, within 3 years after the date of enactment of the Antiterrorism and Effective Death Penalty Act of 1996 [enacted April 24, 1996], will be or is incorporated in a military device within the territory of the United States and remains an integral part of such military device, or is intended to be, or is incorporated in, and remains an integral part of a military device that is intended to become, or has become, the property of any agency of the United States performing military or police functions (including any military reserve component) or the National Guard of any State, wherever such device is located.

(3) For purposes of this subsection, the term "military device" includes, but is not restricted to, shells, bombs, projectiles, mines, missiles, rockets, shaped charges, grenades, perforators, and similar devices lawfully manufactured exclusively for military or police purposes.

(Added Pub. L. 91-452, title XI, Sec. 1102(a), Oct. 15. 1970, 84 Stat. 958; amended Pub. L. 93-639, Sec. 101, Jan. 4, 1975, 88 Stat. 2217; Pub. L. 104-132, title VI, Sec. 605, Apr. 24, 1996, 110 Stat. 1289; Pub. L. 107-296, title XI, Sec. 1126, Nov. 25, 2002, 116 Stat. 2285; Pub. L. 109-59, title VII, Sec. 7127, Aug. 10, 2005, 119 Stat. 1909.)

§ 846. Additional powers of the Attorney General

(a) The Attorney General is authorized to inspect the site of any accident, or fire, in which there is reason to believe that explosive materials were involved, in order that if any such incident has been brought about by accidental means, precautions may be taken to prevent similar accidents from occurring. In order to carry out the purpose of this subsection, the Attorney General is authorized to enter into or upon any property where explosive materials have been used, are suspected of having been used, or have been found in an otherwise unauthorized location. Nothing in this chapter shall be construed as modifying or otherwise affecting in any way the investigative authority of any other Federal agency. In addition to any other investigatory authority they have with respect to violations of provisions of this, the Federal Bureau of Investigation, together with the Bureau of Alcohol, Tobacco, Firearms, and Explosives, shall have authority to conduct investigations with respect to violations of subsection (d), (e), (f), (g), (h), or (i) of section 844 of this title.

(b) The Attorney General is authorized to establish a national repository of information on incidents involving arson and the suspected criminal misuse of explosives. All Federal agencies having information concerning such incidents shall report the information to the Attorney General pursuant to such regulations as deemed necessary to carry out the provisions of this subsection. The repository shall also contain information on incidents voluntarily reported to the Attorney General by State and local authorities.

(Added Pub. L. 91-452, title XI, Sec. 1102(a), Oct. 15, 1970, 84 Stat. 959; amended Pub. L. 104-208, div. A, title I, Sec. 101(f) (title VI, Sec. 654(a)), Sept. 30, 1996, 110 Stat. 3009-314, 3009-369; Pub. L. 107-296, title XI, Sec. 1112(e)(2), (3), Nov. 25, 2002, 116 Stat. 2276.)

§ 847. Rules and regulations

The administration of this chapter shall be vested in the Attorney General. The Attorney General may prescribe such rules and regulations as he deems reasonably necessary to carry out the provisions of this chapter. The Attorney General shall give reasonable public notice, and afford to interested parties opportunity for hearing, prior to prescribing such rules and regulations.

(Added Pub. L. 91-452, title XI, Sec. 1102(a), Oct. 15, 1970, 84 Stat. 959; amended Pub. L. 107-296, title XI, Sec. 1112(e)(3), Nov. 25, 2002, 116 Stat. 2276.)

§ 848. Effect on State Law

No provision of this chapter shall be construed as indicating an intent on the part of the Congress to occupy the field in which such provision operates to the exclusion of the law of any State on the same subject matter, unless there is a direct and positive conflict between such provision and the law of the State so that the two cannot be reconciled or consistently stand together.

(Added Pub. L. 91-452, title XI, Sec. 1102(a), Oct. 15, 1970, 84 Stat. 959.)

Title 27, Code of Federal Regulations
Part 555—Commerce in Explosives

Section Contents

Subpart H—Exemptions

Subpart I—Unlawful Acts, Penalties, Seizures and Forfeitures

Subpart J—Marking of Plastic Explosives

Subpart K—Storage

Authority: 18 U.S.C. 847.

Source: T.D. ATF–87, 46 FR 40384, Aug. 7, 1981, unless otherwise noted.

Editor's Note: There are several name changes for titles referenced in the regulations, 27 CFR Part 555. The title "Chief, Firearms and Explosives Licensing Center" has been changed to "Chief, National Explosives Licensing Center". Also, the title "Regional director (compliance)" has been changed to "Director, Industry Operations" within the respective ATF Field Divisions. All regulations and rulings contained within this publication were current as of 1/26/07. For more recent information, please refer to the ATF website at www.atf.gov.

Subpart A—Introduction

§ 555.1 Scope of regulations.

(a) In general. The regulations contained in this part relate to commerce in explosives and implement Title XI, Regulation of Explosives (18 U.S.C. Chapter 40; 84 Stat. 952), of the Organized Crime Control Act of 1970 (84 Stat. 922), Pub. L. 103–322 (108 Stat. 1796), Pub. L. 104–132 (110 Stat. 1214), and Pub. L. 107–296 (116 Stat. 2135).

(b) Procedural and substantive requirements. This part contains the procedural and substantive requirements relative to:

(1) The interstate or foreign commerce in explosive materials;

(2) The licensing of manufacturers and importers of, and dealers in, explosive materials;

(3) The issuance of permits;

(4) The conduct of business by licensees and operations by permittees;

(5) The storage of explosive materials;

(6) The records and reports required of licensees and permittees;

(7) Relief from disabilities under this part;

(8) Exemptions, unlawful acts, penalties, seizures, and forfeitures; and

(9) The marking of plastic explosives.

[T.D. ATF–87, 46 FR 40384, Aug. 7, 1981, as amended by T.D. ATF–363, 60 FR 17449, Apr. 6, 1995; T.D. ATF–387, 62 FR 8376, Feb. 25, 1997; ATF No. 1, 68 FR 13780, Mar. 20, 2003]

§ 555.2 Relation to other provisions of law.

The provisions in this part are in addition to, and are not in lieu of, any other provision of law, or regulations, respecting commerce in explosive materials. For regulations applicable to commerce in firearms and ammunition, see Part 478 of this chapter. For regulations applicable to traffic in machine guns, destructive devices, and certain other firearms, see Part 479 of this chapter. For statutes applicable to the registration and licensing of persons engaged in the business of manufacturing, importing or exporting arms, ammunition, or implements of war, see section 38 of the Arms Export Control Act (22 U.S.C. 2778), and regulations of Part 447 of this chapter and in Parts 121 through 128 of Title 22, Code of Federal Regulations. For statutes applicable to nonmailable materials, see 18 U.S.C. 1716 and implementing regulations. For statutes applicable to water quality standards, see 33 U.S.C. 1341.

Subpart B—Definitions

§ 555.11 Meaning of terms.

When used in this part, terms are defined as follows in this section. Words in the plural form include the singular, and vice versa, and words indicating the masculine gender include the feminine. The terms "includes" and "including" do not exclude other things not named which are in the same general class or are otherwise within the scope of the term defined.

Act. 18 U.S.C. Chapter 40.

Adjudicated as a mental defective. (a) A determination by a court, board, commission, or other lawful authority that a person, as a result of marked subnormal intelligence, or mental illness, incompetency, condition, or disease:

(1) Is a danger to himself or to others; or

(2) Lacks the mental capacity to contract or manage his own affairs.

(b) The term will include—

(1) A finding of insanity by a court in a criminal case; and

(2) Those persons found incompetent to stand trial or found not guilty by reason of lack of mental responsibility by any court or pursuant to articles 50a and 76b of the Uniform Code of Military Justice, 10 U.S.C. 850a, 876b.

Alien. Any person who is not a citizen or national of the United States.

Ammunition. Small arms ammunition or cartridge cases, primers, bullets, or smokeless propellants designed for use in small arms, including percussion caps, and 3/32 inch and other external burning pyrotechnic hobby fuses. The term does not include black powder.

Appropriate identifying information. The term means, in relation to an individual:

(a) The full name, date of birth, place of birth, sex, race, street address, State of residence, telephone numbers (home and work), country or countries of citizenship, and position at the employer's business or operations of responsible persons and employees authorized to possess explosive materials;

(b) The business name, address, and license or permit number with which the responsible person or employee is affiliated;

(c) If an alien, INS-issued alien number or admission number; and

(d) Social security number, as optional information (this information is not required but is helpful in avoiding misidentification when a background check is conducted).

Approved storage facility. A place where explosive materials are stored, consisting of one or more approved magazines, conforming to the requirements of this part and covered by a license or permit issued under this part.

Articles pyrotechnic. Pyrotechnic devices for professional use similar to consumer fireworks in chemical composition and construction but not intended for consumer use. Such articles meeting the weight limits for consumer fireworks but not labeled as such and classified by U.S. Department of Transportation regulations in 49 CFR 172.101 as UN0431 or UN0432.

Artificial barricade. An artificial mound or revetted wall of earth of a minimum thickness of three feet, or any other approved barricade that offers equivalent protection.

ATF. (a) Prior to January 24, 2003. The Bureau of Alcohol, Tobacco and Firearms, Department of the Treasury, Washington, DC.

(b) On and after January 24, 2003. The Bureau of Alcohol, Tobacco, Firearms and Explosives, Department of Justice, Washington, DC.

ATF officer. (a) Prior to January 24, 2003. An officer or employee of the Bureau of Alcohol, Tobacco and Firearms (ATF) authorized to perform any function relating to the administration or enforcement of this part.

(b) On and after January 24, 2003. An officer or employee of the Bureau of Alcohol, Tobacco, Firearms and Explosives (ATF) authorized to perform any function relating to the administration or enforcement of this part.

Authority having jurisdiction for fire safety. The fire department having jurisdiction over sites where explosives are manufactured or stored.

Barricaded. The effective screening of a magazine containing explosive materials from another magazine, a building, a railway, or a highway, either by a natural barricade or by an artificial barricade. To be properly barricaded, a straight line from the top of any sidewall of the magazine containing explosive materials to the eave line of any other magazine or building, or to a point 12 feet above the center of a railway or highway, will pass through the natural or artificial barricade.

Blasting agent. Any material or mixture, consisting of fuel and oxidizer, that is intended for blasting and not otherwise defined as an explosive; if the finished product, as mixed for use or shipment, cannot be detonated by means of a number 8 test blasting cap when unconfined. A number 8 test blasting cap is one containing 2 grams of a mixture of 80 percent mercury fulminate and 20 percent potassium chlorate, or a blasting cap of equivalent strength. An equivalent strength cap comprises 0.40–0.45 grams of PETN base charge pressed in an aluminum shell with bottom thickness not to exceed to 0.03 of an inch, to a specific gravity of not less than 1.4 g/cc., and primed with standard weights of primer depending on the manufacturer.

Bulk salutes. Salute components prior to final assembly into aerial shells, and finished salute shells held separately prior to being packed with other types of display fireworks.

Bullet-sensitive explosive materials. Explosive materials that can be exploded by 150-grain M2 ball ammunition having a nominal muzzle velocity of 2700 fps (824 mps) when fired from a .30 caliber rifle at a distance of 100 ft (30.5 m), measured perpendicular. The test material is at a temperature of 70 to 75 degrees F (21 to 24 degrees C) and is placed against a ½ inch (12.4 mm) steel backing plate.

Bureau. (a) Prior to January 24, 2003. The Bureau of Alcohol, Tobacco and Firearms, Department of the Treasury, Washington, DC.

(b) On and after January 24, 2003. The Bureau of Alcohol, Tobacco, Firearms and Explosives, Department of Justice, Washington, DC.

Business premises. When used with respect to a manufacturer, importer, or dealer, the property on which explosive materials are manufactured, imported, stored or distributed. The premises include the property where the records of a manufacturer, importer, or dealer are kept if different than the premises where explosive materials are manufactured, imported, stored or distributed. When used with respect to a user of explosive materials, the property on which the explosive materials are received or stored. The premises includes the property where the records of the users are kept if different than the premises where explosive materials are received or stored.

Chief, Firearms and Explosives Licensing Center. The ATF official responsible for the issuance and renewal of licenses and permits under this part.

Committed to a mental institution. A formal commitment of a person to a mental institution by a court, board, commission, or other lawful authority. The term includes a commitment to a mental institution involuntarily. The term includes commitment for mental defectiveness or mental illness. It also includes commitments for other reasons, such as for drug use. The term does not include a person in a mental institution for observation or a voluntary admission to a mental institution.

Common or contract carrier. Any individual or organization engaged in the business of transporting passengers or goods.

Consumer fireworks. Any small firework device designed to produce visible effects by combustion and which must comply with the construction, chemical composition, and labeling regulations of the U.S. Consumer Product Safety Commission, as set forth in title 16, Code of Federal Regulations, parts 1500 and 1507. Some small devices designed to produce audible effects are included, such as whistling devices, ground devices containing 50 mg or less of explosive materials, and aerial devices containing 130 mg or less of explosive materials. Consumer fireworks are classified as fireworks UN0336, and UN0337 by the U.S. Department of Transportation at 49 CFR 172.101. This term does not include fused setpieces containing components which together exceed 50 mg of salute powder.

Controlled substance. A drug or other substance, or immediate precursor, as defined in section 102 of the Controlled Substances Act, 21 U.S.C. 802. The term includes, but is not limited to, marijuana, depressants, stimulants, and narcotic drugs. The term does not include distilled spirits, wine, malt beverages, or tobacco, as those terms are defined or used in Subtitle E of the Internal Revenue Code of 1986, as amended.

Crime punishable by imprisonment for a term exceeding one year. Any offense for which the maximum penalty, whether or not imposed, is capital punishment or imprisonment in excess of one year. The term does not include (a) any Federal or State offenses pertaining to antitrust violations, unfair trade practices, restraints of trade, or (b) any State offense (other than one involving a firearm or explosive) classified by the laws of the State as a misdemeanor and punishable by a term of imprisonment of two years or less.

Customs officer. Any officer of the Customs Service or any commissioned, warrant, or petty officer of the Coast Guard, or any agent or other person authorized to perform the duties of an officer of the Customs Service.

Dealer. Any person engaged in the business of distributing explosive materials at wholesale or retail.

Detonator. Any device containing a detonating charge that is used for initiating detonation in an explosive. The term includes, but is not limited to, electric blasting caps of instantaneous and delay types, blasting caps for use with safety fuses, detonating-cord delay connectors, and nonelectric instantaneous and delay blasting caps.

Director. (a) Prior to January 24, 2003. The Director, Bureau of Alcohol, Tobacco and Firearms, Department of the Treasury, Washington, DC. (b) On and after January 24, 2003. The Director, Bureau of Alcohol, Tobacco, Firearms and Explosives, Department of Justice, Washington, DC.

Discharged under dishonorable conditions. Separation from the U.S. Armed Forces resulting from a dishonorable discharge or dismissal adjudged by general court-martial. The term does not include any separation from the Armed Forces resulting from any other discharge, e.g., a bad conduct discharge.

Display fireworks. Large fireworks designed primarily to produce visible or audible effects by combustion, deflagration, or detonation. This term includes, but is not limited to, salutes containing more than 2 grains (130 mg) of explosive materials, aerial shells containing more than 40 grams of pyrotechnic compositions, and other display pieces which exceed the limits of explosive materials for classification as "consumer fireworks." Display fireworks are classified as fireworks UN0333, UN0334 or UN0335 by the U.S. Department of Transportation at 49 CFR 172.101. This term also includes fused setpieces containing components which together exceed 50 mg of salute powder.

Distribute. To sell, issue, give, transfer, or otherwise dispose of. The term does not include a mere change of possession from a person to his agent or employee in connection with the agency or employment.

Executed under penalties of perjury. Signed with the required declaration under the penalties of perjury as provided on or with respect to the return, form, or other document or, where no form of declaration is required, with the declaration:

"I declare under the penalties of perjury that this —(insert type of document, such as, statement, application, request, certificate), including the documents submitted in support thereof, has been examined by me and, to the best of my knowledge and belief, is true, correct, and complete".

Explosive actuated device. Any tool or special mechanized device which is actuated by explosives, but not a propellent actuated device.

Explosive materials. Explosives, blasting agents, water gels and detonators. Explosive materials include, but are not limited to, all items in the "List of Explosive Materials" provided for in §555.23.

Explosives. Any chemical compound, mixture, or device, the primary or common purpose of which is to function by explosion. The term includes, but is not limited to, dynamite and other high explosives, black powder, pellet powder, initiating explosives, detonators, safety fuses, squibs, detonating cord, igniter cord, and igniters.

Fireworks. Any composition or device designed to produce a visible or an audible effect by combustion, deflagration, or detonation, and which meets the definition of "consumer fireworks" or "display fireworks" as defined by this section.

Fireworks mixing building. Any building or area used for mixing and blending pyrotechnic compositions except wet sparkler mix.

Fireworks nonprocess building. Any office building or other building or area in a fireworks plant where no fireworks, pyrotechnic compositions or explosive materials are processed or stored.

Fireworks plant. All land and buildings thereon used for or in connection with the assembly or processing of fireworks, including warehouses used with or in connection with fireworks plant operations.

Fireworks plant warehouse. Any building or structure used exclusively for the storage of materials which are neither explosive materials nor pyrotechnic compositions used to manufacture or assemble fireworks.

Fireworks process building. Any mixing building; any building in which pyrotechnic compositions or explosive materials is pressed or otherwise prepared for finished and assembly; or any finishing or assembly building.

Fireworks shipping building. A building used for the packing of assorted display fireworks into shipping cartons for individual public displays and for the loading of packaged displays for shipment to purchasers.

Flash powder. An explosive material intended to produce an audible report and a flash of light when ignited which includes but is not limited to oxidizers such as potassium chlorate or potassium perchlorate, and fuels such as sulfur or aluminum powder.

Fugitive from justice. Any person who has fled from the jurisdiction of any court of record to avoid prosecution for any crime or to avoid giving testimony in any criminal proceeding. The term also includes any person who has been convicted of any crime and has fled to avoid imprisonment.

Hardwood. Oak, maple, ash, hickory, or other hard wood, free from loose knots, spaces, or similar defects.

Highway. Any public street, public alley, or public road, including a privately financed, constructed, or maintained road that is regularly and openly traveled by the general public.

Identification document. A document containing the name, residence address, date of birth, and photograph of the holder and which was made or issued by or under the authority of the United States Government, a State, political subdivision of a State, a foreign government, a political subdivision of a foreign government, an international governmental or an international quasi-governmental organization which, when completed with information concerning a particular individual, is of a type intended or commonly accepted for the purpose of identification of individuals.

Importer. Any person engaged in the business of importing or bringing explosive materials into the United States for purposes of sale or distribution.

Indictment. Includes an indictment or information in any court under which a crime punishable by imprisonment for a term exceeding one year may be prosecuted.

Inhabited building. Any building regularly occupied in whole or in part as a habitation for human beings, or any church, schoolhouse, railroad station, store, or other structure where people are accustomed to assemble, except any building occupied in connection with the manufacture, transportation, storage, or use of explosive materials.

Interstate or foreign commerce. Commerce between any place in a State and any place outside of that State, or within any possession of the United States or the District of Columbia, and commerce between places within the same State but through any place outside of that State.

Licensed dealer. A dealer licensed under this part.

Licensed importer. An importer licensed under this part.

Licensed manufacturer. A manufacturer licensed under this part to engage in the business of manufacturing explosive materials for purposes of sale or distribution or for his own use.

Licensee. Any importer, manufacturer, or dealer licensed under this part.

Limited permit. A permit issued to a person authorizing him to receive for his use explosive materials from a licensee or permittee in his state of residence on no more than 6 occasions during the 12-month period in which the permit is valid. A limited permit does not authorize the receipt or transportation of explosive materials in interstate or foreign commerce.

Magazine. Any building or structure, other than an explosives manufacturing building, used for storage of explosive materials.

Manufacturer. Any person engaged in the business of manufacturing explosive materials for purposes of sale or distribution or for his own use.

Mass detonation (mass explosion). Explosive materials mass detonate (mass explode) when a unit or any part of a larger quantity of explosive material explodes and causes all or a substantial part of the remaining material to detonate or explode.

Mental institution. Includes mental health facilities, mental hospitals, sanitariums, psychiatric facilities, and other facilities that provide diagnoses by licensed professionals of mental retardation or mental illness, including a psychiatric ward in a general hospital.

Natural barricade. Natural features of the ground, such as hills, or timber of sufficient density that the surrounding exposures which require protection cannot be seen from the magazine when the trees are bare of leaves.

Number 8 test blasting cap. (See definition of "blasting agent.")

Permittee. Any user of explosives for a lawful purpose who has obtained either a user permit or a limited permit under this part.

Person. Any individual, corporation, company, association, firm, partnership, society, or joint stock company.

Plywood. Exterior, construction grade (laminated wood) plywood.

Propellant actuated device. Any tool or special mechanized device or gas generator system which is actuated by a propellant or which releases and directs work through a propellant charge.

Pyrotechnic compositions. A chemical mixture which, upon burning and without explosion, produces visible, brilliant displays, bright lights, or sounds.

Railway. Any steam, electric, or other railroad or railway which carries passengers for hire.

Region. A geographical region of the Bureau of Alcohol, Tobacco and Firearms.

Regional director (compliance). The principal regional official responsible for administering regulations in this part.

Renounced U.S. citizenship. (a) A person has renounced his U.S. citizenship if the person, having been a citizen of the United States, has renounced citizenship either—

(1) Before a diplomatic or consular officer of the United States in a foreign state pursuant to 8 U.S.C. 1481(a)(5); or

(2) Before an officer designated by the Attorney General when the United States is in a state of war pursuant to 8 U.S.C. 1481(a)(6).

(b) The term will not include any renunciation of citizenship that has been reversed as a result of administrative or judicial appeal.

Responsible person. An individual who has the power to direct the management and policies of the applicant pertaining to explosive materials. Generally, the term includes partners, sole proprietors, site managers, corporate officers and directors, and majority shareholders.

Salute. An aerial shell, classified as a display firework, that contains a charge of flash powder and is designed to produce a flash of light and a loud report as the pyrotechnic effect.

Screen barricade. Any barrier that will contain the embers and debris from a fire or deflagration in a process building, thus preventing propagation of fire to other buildings or areas. Such barriers shall be constructed of metal roofing, ¼ to ½ inch (6 to 13 mm) mesh screen, or equivalent material. The barrier extends from floor level to a height such that a straight line from the top of any side wall of the donor building to the eave line of any exposed building intercepts the screen at a point not less than 5 feet (1.5 m) from the top of the screen. The top 5 feet (1.5 m) of the screen is inclined towards the donor building at an angle of 30 to 45 degrees.

Softwood. Fir, pine, or other soft wood, free from loose knots, spaces, or similar defects.

State. A State of the United States. The term includes the District of Columbia, the Commonwealth of Puerto Rico, and the possessions of the United States.

State of residence. The State in which an individual regularly resides or maintains his home. Temporary stay in a State does not make the State of temporary stay the State of residence.

Theatrical flash powder. Flash powder commercially manufactured in premeasured kits not exceeding 1 ounce and mixed immediately prior to use and intended for use in theatrical shows, stage plays, band concerts, magic acts, thrill shows, and clown acts in circuses.

Unlawful user of or addicted to any controlled substance. A person who uses a controlled substance and has lost the power of self-control with reference to the use of a controlled substance; and any person who is a current user of a controlled substance in a manner other than as prescribed by a licensed physician. Such use is not limited to the use of drugs on a particular day, or within a matter of days or weeks before possession of the explosive materials, but rather that the unlawful use has occurred recently enough to indicate that the individual is actively engaged in such conduct. A person may be an unlawful current user of a controlled substance even though the substance is not being used at the precise time the person seeks to acquire explosive materials or receives or possesses explosive materials. An inference of current use may be drawn from evidence of a recent use or possession of a controlled substance or a pattern of use or possession that reasonably covers the present time, e.g., a conviction for use or possession of a controlled substance within the past year; multiple arrests for such offenses within the past 5 years if the most recent arrest occurred within the past year; or persons found through a drug test to use a controlled substance unlawfully, provided that the test was administered within the past year. For a current or former member of the Armed Forces, an inference of current use may be drawn from recent disciplinary or other administrative action based on confirmed drug use, e.g., court-martial conviction, nonjudicial punishment, or an administrative discharge based on drug use or drug rehabilitation failure.

U.S.C. The United States Code.

User-limited permit. A user permit valid only for a single purchase transaction, a new permit being required for a subsequent purchase transaction.

User permit. A permit issued to a person authorizing him (a) to acquire for his own use explosive materials from a licensee in a State other than the State in which he resides or from a foreign country, and (b) to transport explosive materials in interstate or foreign commerce.

Water gels. Explosives or blasting agents that contain a substantial proportion of water.

(18 U.S.C. 847 (84 Stat. 959); 18 U.S.C. 926 (82 Stat. 1226)
[T.D. ATF–87, 46 FR 40384, Aug. 7, 1981, as amended by T.D. ATF–92, 46 FR 46916, Sept. 23, 1981; T.D. ATF–200, 50 FR 10497, Mar. 15, 1985; T.D. ATF–290, 54 FR 53053, Dec. 27, 1989; T.D. ATF–293, 55 FR 3720, Feb. 5, 1990; T.D. ATF–314, 56 FR 49140, Sept. 27, 1991; T.D. ATF–382, 61 FR 38084, July 23, 1996; T.D. ATF–400, 63 FR 45001, Aug. 24, 1998; T.D. ATF–446, 66 FR 16602, Mar. 27, 2001; ATF No. 1, 68 FR 13780, Mar. 20, 2003]

Subpart C—Administrative and Miscellaneous Provisions

§ 555.21 Forms prescribed.

(a) The Director is authorized to prescribe all forms required by this part. All of the information called for in each form shall be furnished as indicated by the headings on the form and the instructions on or pertaining to the form. In addition, information called for in each form shall be furnished as required by this part.

(b) Requests for forms should be mailed to the ATF Distribution Center, 7943 Angus Court, Springfield, Virginia 22153.

[T.D. ATF–92, 46 FR 46916, Sept. 23, 1981, as amended by T.D. ATF–249, 52 FR 5961, Feb. 27, 1987; T.D. 372, 61 FR 20724, May 8, 1996]

§ 555.22 Alternate methods or procedures; emergency variations from requirements.

(a) Alternate methods or procedures. The permittee or licensee, on specific approval by the Director as provided by this paragraph, may use an alternate method or procedure in lieu of a method or procedure specifically prescribed in this part. The Director may approve an alternate method or procedure, subject to stated conditions, when he finds that:

(1) Good cause is shown for the use of the alternate method or procedure;

(2) The alternate method or procedure is within the purpose of, and consistent with the effect intended by, the specifically prescribed method or procedure and that the alternate method or procedure is substantially equivalent to that specifically prescribed method or procedure; and

(3) The alternate method or procedure will not be contrary to any provision of law and will not result in an increase in cost to the Government or hinder the effective administration of this part. Where the permittee or licensee desires to employ an alternate method or procedure, he shall submit a written application to the regional director (compliance), for transmittal to the Director. The application shall specifically describe the proposed alternate method or procedure and shall set forth the reasons for it. Alternate methods or procedures may not be employed until the application is approved by the Director. The permittee or licensee shall, during the period of authorization of an alternate method or procedure, comply with the terms of the approved application. Authorization of any alternate method or procedure may be withdrawn whenever, in the judgment of the Director, the effective administration of this part is hindered by the continuation of the authorization. As used in this paragraph, alternate methods or procedures include alternate construction or equipment.

(b) Emergency variations from requirements. The Director may approve construction, equipment, and methods of operation other than as specified in this part, where he finds that an emergency exists and the proposed variations from the specified requirements are necessary and the proposed variations:

(1) Will afford security and protection that are substantially equivalent to those prescribed in this part;

(2) Will not hinder the effective administration of this part; and

(3) Will not be contrary to any provisions of law.

Variations from requirements granted under this paragraph are conditioned on compliance with the procedures, conditions, and limitations set forth in the approval of the application. Failure to comply in good faith with the procedures, conditions, and limitations shall automatically terminate the authority for the variations and the licensee or permittee shall fully comply with the prescribed requirements of regulations from which the variations were authorized. Authority for any variation may be withdrawn whenever, in the judgment of the Director, the effective administration of this part is hindered by the continuation of the variation. Where the licensee or permittee desires to employ an emergency variation, he shall submit a written application to the regional director (compliance) for transmittal to the Director. The application shall describe the proposed variation and set forth the reasons for it. Variations may not be employed until the application is approved, except when the emergency requires immediate action to correct a situation that is threatening to life or property. Corrective action may then be taken concurrent with the filing of the application and notification of the Director via telephone.

(c) Retention of approved variations. The licensee or permittee shall retain, as part of his records available for examination by ATF officers, any application approved by the Director under this section.

§ 555.23 List of explosive materials.

The Director shall compile a list of explosive materials, which shall be published and revised at least annually in the Federal Register. The "List of Explosive Materials" (ATF Publication 5400.8) is available at no cost upon request from the ATF Distribution Center, 7943 Angus Court, Springfield, Virginia 22153.

[T.D. ATF–290, 54 FR 53054, Dec. 27, 1989, as amended by T.D. ATF–446, 66 FR 16602, Mar. 27, 2001]

§ 555.24 Right of entry and examination.

(a) Any ATF officer may enter during business hours the premises, including places of storage, of any licensee or holder of a user permit for the purpose of inspecting or examining any records or documents required to be kept under this part, and any facilities in which explosive materials are kept or stored.

(b) Any ATF officer may inspect the places of storage for explosive materials of an applicant for a limited permit or, in the case of a holder of a limited permit, at the time of renewal of such permit.

(c) The provisions of paragraph (b) of this section do not apply to an applicant for the renewal of a limited permit if an ATF officer has, within the preceding 3 years, verified by inspection that the applicant's place of storage for explosive materials meets the requirements of subpart K of this part.
[ATF No. 1, 68 FR 13781, Mar. 20, 2003]

§ 555.25 Disclosure of information.

Upon receipt of written request from any State or any political subdivision of a State, the regional director (compliance) may make available to the State or political subdivision any information which the regional director (compliance) may obtain under the Act with respect to the identification of persons within the State or political subdivision, who have purchased or received explosive materials, together with a description of the explosive materials.

§ 555.26 Prohibited shipment, transportation, receipt, possession, or distribution of explosive materials.

(a) General. No person, other than a licensee or permittee knowingly may transport, ship, cause to be transported, or receive any explosive materials: Provided, That the provisions of this paragraph (a) do not apply to the lawful purchase by a nonlicensee or nonpermittee of commercially manufactured black powder in quantities not to exceed 50 pounds, if the black powder is intended to be used solely for sporting, recreational, or cultural purposes in antique firearms as defined in 18 U.S.C. 921(a)(16), or in antique devices as exempted from the term "destructive device" in 18 U.S.C. 921(a)(4).

(b) Holders of a limited permit. No person who is a holder of a limited permit may—

(1) Transport, ship, cause to be transported, or receive in interstate or foreign commerce any explosive materials;

(2) Receive explosive materials from a licensee or permittee, whose premises are located outside the State of residence of the limited permit holder; or

(3) Receive explosive materials on more than 6 separate occasions, during the period of the permit, from one or more licensees or permittees whose premises are located within the State of residence of the limited permit holder. (See §555.105(b) for the definition of "6 separate occasions.")

(c) Possession by prohibited persons. No person may ship or transport any explosive material in or affecting interstate or foreign commerce or receive or possess any explosive materials which have been shipped or transported in or affecting interstate or foreign commerce who:

(1) Is under indictment or information for, or who has been convicted in any court of, a crime punishable by imprisonment for a term exceeding one year;

(2) Is a fugitive from justice;

(3) Is an unlawful user of or addicted to any controlled substance (as defined in section 102 of the Controlled Substances Act (21 U.S.C. 802) and §555.11);

(4) Has been adjudicated as a mental defective or has been committed to a mental institution;

(5) Is an alien, other than an alien who—

(i) Is lawfully admitted for permanent residence (as that term is defined in section 101(a)(20) of the Immigration and Nationality Act (8 U.S.C. 1101)); or

(ii) Is in lawful nonimmigrant status, is a refugee admitted under section 207 of the Immigration and Nationality Act (8 U.S.C. 1157), or is in asylum status under section 208 of the Immigration and Nationality Act (8 U.S.C. 1158), and—

(A) Is a foreign law enforcement officer of a friendly foreign government, as determined by the Attorney General in consultation with the Secretary of State, entering the United States on official law enforcement business, and the shipping, transporting, possession, or receipt of explosive materials is in furtherance of this official law enforcement business;

(B) Is a person having the power to direct or cause the direction of the management and policies of a corporation, partnership, or association licensed pursuant to section 843(a) of the Act, and the shipping, transporting, possession, or receipt of explosive materials is in furtherance of such power;

(C) Is a member of a North Atlantic Treaty Organization (NATO) or other friendly foreign military force, as determined by the Attorney General in consultation with the Secretary of Defense, (whether or not admitted in a nonimmigrant status) who is present in the United States under military orders for training or other military purpose authorized by the United States, and the shipping, transporting, possession, or receipt of explosive materials is in furtherance of the military purpose; or

(D) Is lawfully present in the United States in cooperation with the Director of Central Intelligence, and the shipment, transportation, receipt, or possession of the explosive materials is in furtherance of such cooperation;

(6) Has been discharged from the armed forces under dishonorable conditions; or

(7) Having been a citizen of the United States, has renounced citizenship.

(d) Distribution to prohibited persons. No person may knowingly distribute explosive materials to any individual who:

(1) Is under twenty-one years of age;

(2) Is under indictment or information for, or who has been convicted in any court of, a crime punishable by imprisonment for a term exceeding one year;

(3) Is a fugitive from justice;

(4) Is an unlawful user of or addicted to any controlled substance (as defined in section 102 of the Controlled Substances Act (21 U.S.C. 802) and §555.11);

(5) Has been adjudicated as a mental defective or has been committed to a mental institution;

(6) Is an alien, other than an alien who—

(i) Is lawfully admitted for permanent residence (as that term is defined in section 101(a)(20) of the Immigration and Nationality Act (8 U.S.C. 1101)); or

(ii) Is in lawful nonimmigrant status, is a refugee admitted under section 207 of the Immigration and Nationality Act (8 U.S.C. 1157), or is in asylum status under section 208 of the Immigration and Nationality Act (8 U.S.C. 1158), and—

(A) Is a foreign law enforcement officer of a friendly foreign government, as determined by the Attorney General in consultation with the Secretary of State, entering the United States on official law enforcement business, and the shipping, transporting, possession, or receipt of explosive materials is in furtherance of this official law enforcement business;

(B) Is a person having the power to direct or cause the direction of the management and policies of a corporation, partnership, or association licensed pursuant to section 843(a) of the Act, and the shipping, transporting, possession, or receipt of explosive materials is in furtherance of such power;

(C) Is a member of a North Atlantic Treaty Organization (NATO) or other friendly foreign military force, as determined by the Attorney General in consultation with the Secretary of Defense, (whether or not admitted in a nonimmigrant status) who is present in the United States under military orders for training or other military purpose authorized by the United States, and the shipping, transporting, possession, or receipt of explosive materials is in furtherance of the military purpose; or

(D) Is lawfully present in the United States in cooperation with the Director of Central Intelligence, and the shipment, transportation, receipt, or possession of the explosive materials is in furtherance of such cooperation;

(7) Has been discharged from the armed forces under dishonorable conditions; or

(8) Having been a citizen of the United States, has renounced citizenship.

(e) See §555.180 for regulations concerning the prohibited manufacture, importation, exportation, shipment, transportation, receipt, transfer, or possession of plastic explosives that do not contain a detection agent.

[ATF No. 1, 68 FR 13781, Mar. 20, 2003]

§ 555.27 Out-of-State disposition of explosive materials.

(a) No nonlicensee or nonpermittee may distribute any explosive materials to any other nonlicensee or nonpermittee who the distributor knows or who has reasonable cause to believe does not reside in the State in which the distributor resides.

(b) The provisions of this section do not apply on and after May 24, 2003.

[ATF No. 1, 68 FR 13782, Mar. 20, 2003]

§ 555.28 Stolen explosive materials.

No person shall receive, conceal, transport, ship, store, barter, sell, or dispose of any stolen explosive materials knowing or having reasonable cause to believe that the explosive materials were stolen.

§ 555.29 Unlawful storage.

No person shall store any explosive materials in a manner not in conformity with this part.

§ 555.30 Reporting theft or loss of explosive materials.

(a) Any licensee or permittee who has knowledge of the theft or loss of any explosive materials from his stock shall, within 24 hours of discovery, report the theft or loss by telephoning 1–800–800–3855 (nationwide toll free number) and on ATF F 5400.5 (formerly Form 4712) in accordance with the instructions on the form. Theft or loss of any explosive materials shall also be reported to appropriate local authorities.

(b) Any other person, except a carrier of explosive materials, who has knowledge of the theft or loss of any explosive materials from his stock shall, within 24 hours of discovery, report the theft or loss by telephoning 1–800–800–3855 (nationwide toll free number) and in writing to the nearest ATF office. Theft or loss shall be reported to appropriate local authorities.

(c) Reports of theft or loss of explosive materials under paragraphs (a) and (b) of this section must include the following information, if known:

(1) The manufacturer or brand name.

(2) The manufacturer's marks of identification (date and shift code).

(3) Quantity (applicable quantity units, such as pounds of explosives, number of detonators, etc.).

(4) Description (dynamite, blasting agents, detonators, etc.) and United Nations (UN) identification number, hazard division number, and classification letter, e.g., 1.1D, as classified by the U.S. Department of Transportation at 49 CFR 172.101 and 173.52.

(5) Size (length and diameter).

(d) A carrier of explosive materials who has knowledge of the theft or loss of any explosive materials shall, within 24 hours of discovery, report the theft or loss by telephoning 1–800–800–3855 (nationwide toll free number). Theft or loss shall also be reported to appropriate local authorities. Reports of theft or loss of explosive materials by carriers shall include the following information, if known:

(1) The manufacturer or brand name.

(2) Quantity (applicable quantity units, such as pounds of explosives, number of detonators, etc.).

(3) Description (United Nations (UN) identification number, hazard division number, and classification letter, e.g., 1.1D) as classified by the U.S. Department of Transportation at 49 CFR 172.101 and 173.52.

[T.D. ATF–87, 46 FR 40384, Aug. 7, 1981, as amended by T.D. ATF–400, 63 FR 45002, Aug. 24, 1998]

§ 555.31 Inspection of site accidents or fires; right of entry.

Any ATF officer may inspect the site of any accident or fire in which there is reason to believe that explosive materials were involved. Any ATF officer may enter into or upon any property where explosive materials have been used, are suspected of having been used, or have been found in an otherwise unauthorized location.

§ 555.32 Special explosive devices.

The Director may exempt certain explosive actuated devices, explosive actuated tools, or similar devices from the requirements of this part. A person who desires to obtain an exemption under this section for any special explosive device, which as designed does not constitute a public safety or security hazard, shall submit a written request to the Director. Each request shall be executed under the penalties of perjury and contain a complete and accurate description of the device, the name and address of the manufacturer or importer, the purpose of and use for which it is intended, and any photographs, diagrams, or drawings as may be necessary to enable the Director to make a determination. The Director may require that a sample of the device be submitted for examination and evaluation. If it is not possible to submit the device, the person requesting the exemption shall advise the Director and designate the place where the device will be available for examination and evaluation.

§ 555.33 Background checks and clearances (effective May 24, 2003).

(a) Background checks.

(1) If the Director receives from a licensee or permittee the names and appropriate identifying information of responsible persons and employees who will be authorized by the employer to possess explosive materials in the course of employment with the employer, the Director will conduct a background check in accordance with this section.

(2) The Director will determine whether the responsible person or employee is one of the persons described in any paragraph of section 842(i) of the Act (see §555.26). In making such determination, the Director may take into account a letter or document issued under paragraph (a)(3) of this section.

(3)(i) If the Director determines that the responsible person or the employee is not one of the persons described in any paragraph of section 842(i) of the Act (see §555.26), the Director will notify the employer in writing or electronically of the determination and issue, to the responsible person or employee, as the case may be, a letter of clearance which confirms the determination.

(ii) If the Director determines that the responsible person or employee is one of the persons described in any paragraph of section 842(i) of the Act (see §555.26), ATF will notify the employer in writing or electronically of the determination and issue to the responsible person or the employee, as the case may be, a document that confirms the determination; explains the grounds for the determination; provides information on how the disability may be relieved; and explains how the determination may be appealed. The employer will retain the notification as part of his permanent records in accordance with §555.121. The employer will take immediate steps to remove the responsible person from his position directing the management or policies of the business or operations as they relate to explosive materials or, as the case may be, to remove the employee from a position requiring the possession of explosive materials. Also, if the employer has listed the employee as a person authorized to accept delivery of explosive materials, as specified in §555.103 or §555.105, the employer must remove the employee from such list and immediately, and in no event later than the second business day after such change, notify distributors of such change.

(b) Appeals and correction of erroneous system information—

(1) In general. A responsible person or employee may challenge the adverse determination set out in the letter of denial, in writing and within 45 days of issuance of the determination, by directing his or her challenge to the basis for the adverse determination, or to the accuracy of the record upon which the adverse determination is based, to the Director. The appeal request must include appropriate documentation or record(s) establishing the legal and/or factual basis for the challenge. Any record or document of a court or other government entity or official furnished in support of an appeal must be certified by the court or other government entity or official as a true copy. In the case of an employee, or responsible person who did not submit fingerprints, such appeal must be accompanied by two properly completed FBI Forms FD–258 (fingerprint card). The Director will advise the individual in writing of his decision and the reasons for the decision.

(2) Employees. The letter of denial, among other things, will advise an employee who elects to challenge an adverse determination to submit the fingerprint cards as described above. The employee also will be advised of the agency name and address that originated the record containing the information causing the adverse determination ("originating agency"). At that time, and where appropriate, an employee is encouraged to apply to the originating agency to challenge the accuracy of the record(s) upon which the denial is based. The originating agency may respond to the individual's application by addressing the individual's specific reasons for the challenge, and by indicating whether additional information or documents are required. If the record is corrected as a result of the application to the originating agency, the individual may so notify ATF which will, in turn, verify the record correction with the originating agency and take all necessary steps to contact the agency responsible for the record system and correct the record. The employee may provide to ATF additional and appropriate documentation or record(s) establishing the legal and/or factual basis for the challenge to ATF's decision to uphold the initial denial. If ATF does not receive such additional documentation or record(s) within 45 days of the date of the decision upholding the initial denial, ATF will close the appeal.

(3) Responsible persons. The letter of denial, among other things, will advise a responsible person of the agency name and address which originated the record containing the information causing the adverse determination ("originating agency"). A responsible person who elects to challenge the adverse determination, where appropriate, is encouraged to apply to the originating agency to challenge the accuracy of the record(s) upon which the denial is based. The originating agency may respond to the individual's application by addressing the individual's specific reasons for the challenge, and by indicating whether additional information or documents are required. If the record is corrected as a result of the application to the originating agency, the individual may so notify ATF which will, in turn, verify the record correction with the originating agency and take all necessary steps to contact the agency responsible for the record system and correct the record. A responsible person may provide additional documentation or records as specified for employees in paragraph (b)(2) of this section.

(Approved by the Office of Management and Budget under control number 1140–0081)
[ATF No. 1, 68 FR 13783, Mar. 20, 2003]

§ 555.34 Replacement of stolen or lost ATF Form 5400.30 (Intrastate Purchase of Explosives Coupon (IPEC)).

When any Form 5400.30 is stolen, lost, or destroyed, the person losing possession will, upon discovery of the theft, loss, or destruction, immediately, but in all cases before 24 hours have elapsed since discovery, report the matter to the Director by telephoning 1–888–ATF-BOMB (nationwide toll free number). The report will explain in detail the circumstances of the theft, loss, or destruction and will include all known facts that may serve to identify the document. Upon receipt of the report, the Director will make such investigation as appears appropriate and may issue a duplicate document upon such conditions as the circumstances warrant.

(Approved by the Office of Management and Budget under control number 1140–0077)
[ATF No. 1, 68 FR 13783, Mar. 20, 2003]

Subpart D—Licenses and Permits

§ 555.41 General.

(a) Licenses and permits issued prior to May 24, 2003.

(1) Each person intending to engage in business as an importer or manufacturer of, or a dealer in, explosive materials, including black powder, must, before commencing business, obtain the license required by this subpart for the business to be operated. Each person who intends to acquire for use explosive materials from a licensee in a State other than the State in which he resides, or from a foreign country, or who intends to transport explosive materials in interstate or foreign commerce, must obtain a permit under this subpart; except that it is not necessary to obtain a permit if the user intends to lawfully purchase:

(i) Explosive materials from a licensee in a State contiguous to the user's State of residence and the user's State of residence has enacted legislation, currently in force, specifically authorizing a resident of that State to purchase explosive materials in a contiguous State; or

(ii) Commercially manufactured black powder in quantities not to exceed 50 pounds, intended to be used solely for sporting, recreational, or cultural purposes in antique firearms or in antique devices.

(2) Each person intending to engage in business as an explosive materials importer, manufacturer, or dealer must file an application, with the required fee (see §555.42), with ATF in accordance with the instructions on the form (see §555.45). A license will, subject to law, entitle the licensee to transport, ship, and receive explosive materials in interstate or foreign commerce, and to engage in the business specified by the license, at the location described on the license. A separate license must be obtained for each business premises at which the applicant is to manufacture, import, or distribute explosive materials except under the following circumstances:

(i) A separate license will not be required for storage facilities operated by the licensee as an integral part of one business premises or to cover a location used by the licensee solely for maintaining the records required by this part.

(ii) A separate license will not be required of a licensed manufacturer with respect to his on-site manufacturing.

(iii) It will not be necessary for a licensed importer or a licensed manufacturer (for purposes of sale or distribution) to also obtain a dealer's license in order to engage in business on his licensed premises as a dealer in explosive materials.

(iv) A separate license will not be required of licensed manufacturers with respect to their on-site manufacture of theatrical flash powder.

(3) Except as provided in paragraph (a)(1) of this section, each person intending to acquire explosive materials from a licensee in a State other than a State in which he resides, or from a foreign country, or who intends to transport explosive materials in interstate or foreign commerce, must file an application, with the required fee (see §555.43), with ATF in accordance with the instructions on the form (see §555.45). A permit will, subject to law, entitle the permittee to acquire, transport, ship, and receive in interstate or foreign commerce explosive materials. Only one permit is required under this part.

(b) Licenses and permits issued on and after May 24, 2003—

(1) In general.

(i) Each person intending to engage in business as an importer or manufacturer of, or a dealer in, explosive materials, including black powder, must, before commencing business, obtain the license required by this subpart for the business to be operated.

(ii) Each person who intends to acquire for use explosive materials within the State in which he resides on no more than 6 separate occasions during the 12-month period in which the permit is valid must obtain a limited permit under this subpart. (See §555.105(b) for definition of "6 separate occasions.")

(iii) Each person who intends to acquire for use explosive materials from a licensee or permittee in a State other than the State in which he resides, or from a foreign country, or who intends to transport explosive materials in interstate or foreign commerce, or who intends to acquire for use explosive materials within the State in which he resides on more than 6 separate occasions during a 12-month period, must obtain a user permit under this subpart.

(iv) It is not necessary to obtain a permit if the user intends only to lawfully purchase commercially manufactured black powder in quantities not to exceed 50 pounds, intended to be used solely for sporting, recreational, or cultural purposes in antique firearms or in antique devices.

(2) Importers, manufacturers, and dealers. Each person intending to engage in business as an explosive materials importer, manufacturer, or dealer must file an application, with the required fee (see §555.42), with ATF in accordance with the instructions on the form (see §555.45). A license will, subject to law, entitle the licensee to transport, ship, and receive explosive materials in interstate or foreign commerce, and to engage in the business specified by the license, at the location described on the license. A separate license must be obtained for each business premises at which the applicant is to manufacture, import, or distribute explosive materials except under the following circumstances:

(i) A separate license will not be required for storage facilities operated by the licensee as an integral part of one business premises or to cover a location used by the licensee solely for maintaining the records required by this part.

(ii) A separate license will not be required of a licensed manufacturer with respect to his on-site manufacturing.

(iii) It will not be necessary for a licensed importer or a licensed manufacturer (for purposes of sale or distribution) to also obtain a dealer's license in order to engage in business on his licensed premises as a dealer in explosive materials. No licensee will be required to obtain a user permit to lawfully transport, ship, or receive explosive materials in interstate or foreign commerce.

(iv) A separate license will not be required of licensed manufacturers with respect to their on-site manufacture of theatrical flash powder.

(3) Users of explosive materials.

(i) A limited permit will, subject to law, entitle the holder of such permit to receive for his use explosive materials from a licensee or permittee in his state of residence on no more than 6 separate occasions during the 12-month period in which the permit is valid. A limited permit does not authorize the receipt or transportation of explosive materials in interstate or foreign commerce. Holders of limited permits who need to receive explosive materials on more than 6 separate occasions during a 12-month period must obtain a user permit in accordance with this subpart.

(ii) Each person intending to acquire explosive materials from a licensee in a State other than a State in which he resides, or from a foreign country, or who intends to transport explosive materials in interstate or foreign commerce, must file an application for a user permit, with the required fee (see §555.43), with ATF in accordance with the instructions on the form (see §555.45). A user permit will, subject to law, entitle the permittee to transport, ship, and receive in interstate or foreign commerce explosive materials. Only one user permit per person is required under this part, irrespective of the number of locations relating to explosive materials operated by the holder of the user permit.

(Approved by the Office of Management and Budget under control number 1140–0083)

[ATF No. 1, 68 FR 13783, Mar. 20, 2003, as amended by ATF 5F, 70 FR 30633, May 27, 2005]

§ 555.42 License fees.

(a) Each applicant shall pay a fee for obtaining a three year license, a separate fee being required for each business premises, as follows:

(1) Manufacturer—$200.

(2) Importer—$200.

(3) Dealer—$200.

(b) Each applicant for a renewal of a license shall pay a fee for a three year license as follows:

(1) Manufacturer—$100.

(2) Importer—$100.

(3) Dealer—$100.

[T.D. ATF–400, 63 FR 45002, Aug. 24, 1998]

§ 555.43 Permit fees.

(a) Each applicant must pay a fee for obtaining a permit as follows:

(1) User—$100 for a three-year period.

(2) User-limited (nonrenewable)—$75.

(3) Limited—$25 for a one-year period.

(b) **(1)** Each applicant for renewal of a user permit must pay a fee of $50 for a three-year period.

(2) Each applicant for renewal of a limited permit must pay a fee of $12 for a one-year period.

[ATF No. 1, 68 FR 13785, Mar. 20, 2003]

§ 555.44 License or permit fee not refundable.

No refund of any part of the amount paid as a license or permit fee will be made where the operations of the licensee or permittee are, for any reason, discontinued during the period of an issued license or permit. However, the license or permit fee submitted with an application for a license or permit will be refunded if that application is denied, withdrawn, or abandoned, or if a license is cancelled subsequent to having been issued through administrative error.

§ 555.45 Original license or permit.

(a) Licenses issued prior to May 24, 2003. Any person who intends to engage in business as an explosive materials importer, manufacturer, or dealer, or who has not timely submitted application for renewal of a previous license issued under this part, shall file with ATF an application for License, Explosives, ATF F 5400.13 with ATF in accordance with the instructions on the form. The application must be executed under the penalties of perjury and the penalties imposed by 18 U.S.C. 844(a). The application is to be accompanied by the appropriate fee in the form of a money order or check made payable to the Bureau of Alcohol, Tobacco and Firearms. ATF F 5400.13 may be obtained from any ATF office. The Chief, Firearms and Explosives Licensing Center, will not approve an application postmarked on or after March 20, 2003, unless it is submitted with a Responsible Person Questionnaire, ATF Form 5400.28. Form 5400.28 must be completed in accordance with the instructions on the form.

(b) Permits issued prior to May 24, 2003. Any person, except as provided in §555.41(a), who intends to acquire explosive materials from a licensee in a state other than the State in which that person resides, or from a foreign country, or who intends to transport explosive materials in interstate or foreign commerce, or who has not timely submitted application for renewal of a previous permit issued under this part, shall file an application for Permit, Explosives, ATF F 5400.16 or Permit, User Limited Special Fireworks, ATF F 5400.21 with ATF in accordance with the instructions on the form. The application must be executed under the penalties of perjury and the penalties imposed by 18 U.S.C. 844(a). The application is to be accompanied by the appropriate fee in the form of a money order or check made payable to the Bureau of Alcohol, Tobacco and Firearms. ATF F 5400.16 and ATF F 5400.21 may be obtained from any ATF office. The Chief, Firearms and Explosives Licensing Center, will not approve an application postmarked on or after March 20, 2003, unless it is submitted with a Responsible Person Questionnaire, ATF Form 5400.28. Form 5400.28 must be completed in accordance with the instructions on the form.

(c) Licenses and permits issued on and after May 24, 2003—

(1) License. Any person who intends to engage in the business as an importer of, manufacturer of, or dealer in explosive materials, or who has not timely submitted an application for renewal of a previous license issued under this part, must file an application for License, Explosives, ATF F 5400.13, with ATF in accordance with the instructions on the form. ATF Form 5400.13 may be obtained by contacting any ATF office. The application must:

(i) Be executed under the penalties of perjury and the penalties imposed by 18 U.S.C. 844(a);

(ii) Include appropriate identifying information concerning each responsible person;

(iii) Include a photograph and fingerprints for each responsible person;

(iv) Include the names of and appropriate identifying information regarding all employees who will be authorized by the applicant to possess explosive materials by submitting ATF F 5400.28 for each employee; and

(v) Include the appropriate fee in the form of money order or check made payable to the Bureau of Alcohol, Tobacco, Firearms and Explosives.

(2) User permit and limited permit. Except as provided in 555.41(b)(1)(iv), any person who intends to acquire explosive materials in the State in which that person resides or acquire explosive materials from a licensee or holder of a user permit in a State other than the State in which that person resides, or from a foreign country, or who intends to transport explosive materials in interstate or foreign commerce, or who has not timely submitted an application for renewal of a previous permit issued under this part, must file an application for Permit, Explosives, ATF F 5400.16 or Permit, User Limited Display Fireworks, ATF F 5400.21 with ATF in accordance with the instructions on the form. ATF Form 5400.16 and ATF Form 5400.21 may be obtained by contacting any ATF office. The application must:

(i) Be executed under the penalties of perjury and the penalties imposed by 18 U.S.C. 844(a);

(ii) Include a photograph, fingerprints, and appropriate identifying information for each responsible person;

(iii) Include the names of and appropriate identifying information regarding all employees who will be authorized by the applicant to possess explosive materials by submitting ATF F 5400.28 for each employee; and

(iv) Include the appropriate fee in the form of money order or check made payable to the Bureau of Alcohol, Tobacco, Firearms and Explosives.

(3) The Chief, Firearms and Explosives Licensing Center, will conduct background checks on responsible persons and employees authorized by the applicant to possess explosive materials in accordance with §555.33. If it is determined that any responsible person or employee is described in any paragraph of section 842(i) of the Act, the applicant must submit an amended application indicating removal or reassignment of that person before the license or permit will be issued.

(Approved by the Office of Management and Budget under control number 1140–0083)

(18 U.S.C. 847 (84 Stat. 959); 18 U.S.C. 926 (82 Stat. 1226))
[T.D. ATF–200, 50 FR 10497, Mar. 15, 1985, as amended by T.D. ATF–400, 63 FR 45002, Aug. 24, 1998; ATF No. 1, 68 FR 13785, Mar. 20, 2003]

§ 555.46 Renewal of license or permit.

(a) If a licensee or permittee intends to continue the business or operation described on a license or permit issued under this part after the expiration date of the license or permit, he shall, unless otherwise notified in writing by the Chief, Firearms and Explosives Licensing Center, execute and file prior to the expiration of his license or permit an application for license renewal, ATF F 5400.14 (Part III), or an application for permit renewal, ATF F 5400.15 (Part III), accompanied by the required fee, with ATF in accordance with the instructions on the form. In the event the licensee or permittee does not timely file a renewal application, he shall file an original application as required by §555.45, and obtain the required license or permit in order to continue business or operations.

(b) A user-limited permit is not renewable and is valid for a single purchase transaction. Applications for all user-limited permits must be filed on ATF F 5400.16 or ATF F 5400.21, as required by §555.45.

(18 U.S.C. 847 (84 Stat. 959); 18 U.S.C. 926 (82 Stat. 1226))
[T.D. ATF–87, 46 FR 40384, Aug. 7, 1981, as amended by T.D. ATF–200, 50 FR 10497, Mar. 15, 1985; T.D. ATF–290, 54 FR 53054, Dec. 27, 1989; T.D. ATF–400, 63 FR 45002, Aug. 24, 1998]

§ 555.47 Insufficient fee.

If an application is filed with an insufficient fee, the application and fee submitted will be returned to the applicant.

(18 U.S.C. 847 (84 Stat. 959); 18 U.S.C. 926 (82 Stat. 1226))
[T.D. ATF–200, 50 FR 10498, Mar. 15, 1985]

§ 555.48 Abandoned application.

Upon receipt of an incomplete or improperly executed application, the applicant will be notified of the deficiency in the application. If the application is not corrected and returned within 30 days following the date of notification, the application will be considered as having been abandoned and the license or permit fee returned.

§ 555.49 Issuance of license or permit.

(a) Issuance of license or permit prior to May 24, 2003. (1) The Chief, Firearms and Explosives Licensing Center, will issue a license or permit if—

(i) A properly executed application for the license or permit is received; and

(ii) Through further inquiry or investigation, or otherwise, it is found that the applicant is entitled to the license or permit.

(2) The Chief, Firearms and Explosives Licensing Center, will approve a properly executed application for a license or permit, if:

(i) The applicant is 21 years of age or over;

(ii) The applicant (including, in the case of a corporation, partnership, or association, any individual possessing, directly or indirectly, the power to direct or cause the direction of the management and policies of the corporation, partnership, or association) is not a person to whom distribution of explosive materials is prohibited under the Act;

(iii) The applicant has not willfully violated any provisions of the Act or this part;

(iv) The applicant has not knowingly withheld information or has not made any false or fictitious statement intended or likely to deceive, in connection with his application;

(v) The applicant has in a State, premises from which he conducts business or operations subject to license or permit under the Act or from which he intends to conduct business or operations;

(vi) The applicant has storage for the class (as described in §555.202) of explosive materials described on the application, unless he establishes to the satisfaction of the Chief, Firearms and Explosives Licensing Center, that the business or operations to be conducted will not require the storage of explosive materials;

(vii) The applicant has certified in writing that he is familiar with and understands all published State laws and local ordinances relating to explosive materials for the location in which he intends to do business; and

(viii) The applicant for a license has submitted the certificate required by section 21 of the Federal Water Pollution Control Act, as amended (33 U.S.C. 1341).

(3) The Chief, Firearms and Explosives Licensing Center, will approve or the regional director (compliance) will deny any application for a license or permit within the 45-day period beginning on the date a properly executed application was received. However, when an applicant for license or permit renewal is a person who is, under the provisions of §555.83 or §555.142, conducting business or operations under a previously issued license or permit, action regarding the application will be held in abeyance pending the completion of the proceedings against the applicant's existing license or permit, or renewal application, or final action by the Director on an application for relief submitted under §555.142, as the case may be.

(4) The license or permit and one copy will be forwarded to the applicant, except that in the case of a user-limited permit, the original only will be issued.

(5) Each license or permit will bear a serial number and this number may be assigned to the licensee or permittee to whom issued for as long as he maintains continuity of renewal in the same region.

(b) Issuance of license or permit on and after May 24, 2003.

(1) The Chief, Firearms and Explosives Licensing Center, will issue a license or permit if:

(i) A properly executed application for the license or permit is received; and

(ii) Through further inquiry or investigation, or otherwise, it is found that the applicant is entitled to the license or permit.

(2) The Chief, Firearms and Explosives Licensing Center, will approve a properly executed application for a license or permit, if:

(i) The applicant (or, if the applicant is a corporation, partnership, or association, each responsible person with respect to the applicant) is not a person described in any paragraph of section 842(i) of the Act;

(ii) The applicant has not willfully violated any provisions of the Act or this part;

(iii) The applicant has not knowingly withheld information or has not made any false or fictitious statement intended or likely to deceive, in connection with his application;

(iv) The applicant has in a State, premises from which he conducts business or operations subject to license or permit under the Act or from which he intends to conduct business or operations;

(v) The applicant has storage for the class (as described in §555.202) of explosive materials described on the application;

(vi) The applicant has certified in writing that he is familiar with and understands all published State laws and local ordinances relating to explosive materials for the location in which he intends to do business;

(vii) The applicant for a license has submitted the certificate required by section 21 of the Federal Water Pollution Control Act, as amended (33 U.S.C. 1341);

(viii) None of the employees of the applicant who will be authorized by the applicant to possess explosive materials is a person described in any paragraph of section 842(i) of the Act; and

(ix) In the case of an applicant for a limited permit, the applicant has certified in writing that the applicant will not receive explosive materials on more than 6 separate occasions during the 12-month period for which the limited permit is valid.

(3) The Chief, Firearms and Explosives Licensing Center, will approve or the regional director (compliance) will deny any application for a license or permit within the 90-day period beginning on the date a properly executed application was received. However, when an applicant for license or permit renewal is a person who is, under the provisions of §555.83 or §555.142, conducting business or operations under a previously issued license or permit, action regarding the application will be held in abeyance pending the completion of the proceedings against the applicant's existing license or permit, or renewal application, or final action by the Director on an application for relief submitted under §555.142, as the case may be.

(4) The license or permit and one copy will be forwarded to the applicant, except that in the case of a user-limited permit, the original only will be issued.

(5) Each license or permit will bear a serial number and this number may be assigned to the licensee or permittee to whom issued for as long as he maintains continuity of renewal in the same region.

(Approved by the Office of Management and Budget under control number 1140–0082)

[ATF No. 1, 68 FR 13785, Mar. 20, 2003]

§ 555.50 Correction of error on license or permit.

(a) Upon receipt of a license or permit issued under this part, each licensee or permittee shall examine the license or permit to insure that the information on it is accurate. If the license or permit is incorrect, the licensee or permittee shall return the license or permit to the Chief, Firearms and Explosives Licensing Center, with a statement showing the nature of the error. The Chief, Firearms and Explosives Licensing Center, shall correct the error, if the error was made in his office, and return the license or permit. However, if the error resulted from information contained in the licensee's or permittee's application for the license or permit, the Chief, Firearms and Explosives Licensing Center, shall require the licensee or permittee to file an amended application setting forth the correct information and a statement explaining the error contained in the application. Upon receipt of the amended application and a satisfactory explanation of the error, the Chief, Firearms and Explosives Licensing Center, shall make the correction on the license or permit and return it to the licensee or permittee.

(b) When the Chief, Firearms and Explosives Licensing Center, finds through any means other than notice from the licensee or permittee that an incorrect license or permit has been issued, (1) the Chief, Firearms and Explosives Licensing Center, may require the holder of the incorrect license or permit to return the license or permit for correction, and (2) if the error resulted from information contained in the licensee's or permittee's application for the license or permit, the Chief, Firearms and Explosives Licensing Center, shall require the licensee or permittee to file an amended application setting forth the correct information, and a

statement satisfactorily explaining the error contained in the application. The Chief, Firearms and Explosives Licensing Center, then shall make the correction on the license or permit and return it to the licensee or permittee.
[T.D. ATF–87, 46 FR 40384, Aug. 7, 1981, as amended by T.D. ATF–290, 54 FR 53054, Dec. 27, 1989]

§ 555.51 Duration of license or permit.

(a) Prior to May 24, 2003. An original license or permit is issued for a period of three years. A renewal license or permit is issued for a period of three years. However, a user-limited permit is valid only for a single purchase transaction.

(b) On and after May 24, 2003.

(1) An original license or user permit is issued for a period of three years. A renewal license or user permit is also issued for a period of three years. However, a user-limited permit is valid only for a single purchase transaction.

(2) A limited permit is issued for a period of one year. A renewal limited permit is also issued for a period of one year.
[ATF No. 1, 68 FR 13786, Mar. 20, 2003]

§ 555.52 Limitations on license or permit.

(a) The license covers the business of explosive materials specified in the license at the licensee's business premises (see §555.41(b)).

(b) The permit is valid with respect to the type of operations of explosive materials specified in the permit.
[T.D. ATF–87, 46 FR 40384, Aug. 7, 1981, as amended by T.D. ATF–387, 62 FR 8376, Feb. 25, 1997; ATF 5F, 70 FR 30633, May 27, 2005]

§ 555.53 License and permit not transferable.

Licenses and permits issued under this part are not transferable to another person. In the event of the lease, sale, or other transfer of the business or operations covered by the license or permit, the successor must obtain the license or permit required by this part before commencing business or operations. However, for rules on right of succession, see §555.59.

§ 555.54 Change of address.

(a) During the term of a license or permit, a licensee or permittee may move his business or operations to a new address at which he intends to regularly carry on his business or operations, without procuring a new license or permit. However, in every case, the licensee or permittee shall—

(1) Give notification of the new location of the business or operations to the Chief, Firearms and Explosives Licensing Center at least 10 days before the move; and

(2) Submit the license or permit to the Chief, Firearms and Explosives Licensing Center. The Chief, Firearms and Explosives Licensing Center will issue an amended license or permit, which will contain the new address (and new license or permit number, if any).

(b) Licensees and permittees whose mailing address will change must notify the Chief, Firearms and Explosives Licensing Center, at least 10 days before the change.

(Paragraph (b) approved by the Office of Management and Budget under control number 1140–0080)
[T.D. ATF–87, 46 FR 40384, Aug. 7, 1981, as amended by T.D. ATF–290, 54 FR 53054, Dec. 27, 1989; ATF No. 1, 68 FR 13786, Mar. 20, 2003]

§ 555.56 Change in trade name.

A licensee or permittee continuing to conduct business or operations at the location shown on his license or permit is not required to obtain a new license or permit by reason of a mere change in trade name under which he conducts his business or operations. However, the licensee or permittee shall furnish his license or permit and any copies furnished with the license or permit for endorsement of the change to the Chief, Firearms and Explosives Licensing Center, within 30 days from the date the licensee or permittee begins his business or operations under the new trade name.
[T.D. ATF–87, 46 FR 40384, Aug. 7, 1981, as amended by T.D. ATF–290, 54 FR 53054, Dec. 27, 1989]

§ 555.57 Change of control, change in responsible persons, and change of employees.

(a) In the case of a corporation or association holding a license or permit under this part, if actual or legal control of the corporation or association changes, directly or indirectly, whether by reason of change in stock ownership or control (in the corporation holding a license or permit or in any other corporation), by operation of law, or in any other manner, the licensee or permittee shall, within 30 days of the change, give written notification executed under the penalties of perjury, to the Chief, Firearms and Explosives Licensing Center. Upon expiration of the license or permit, the corporation or association shall file an ATF F 5400.13 or an ATF F 5400.16 as required by §555.45, and pay the fee prescribed in §555.42(b) or §555.43(b).

(b) For all licenses or permits issued on and after May 24, 2003, each person holding the license or permit must report to the Chief, Firearms and Explosives Licensing Center, any change in responsible persons or employees authorized to possess explosive materials. Such report must be submitted within 30 days of the change and must include appropriate identifying information for each responsible person. Reports relating to newly hired employees authorized to possess explosive materials must be submitted on ATF F 5400.28 for each employee.

(c) Upon receipt of a report, the Chief, Firearms and Explosives Licensing Center, will conduct a background check, if appropriate, in accordance with §555.33.

(d) The reports required by paragraph (b) of this section must be retained as part of a licensee's or permittee's permanent records for the period specified in §555.121.

(Approved by the Office of Management and Budget under control number 1140–0074)
[T.D. ATF–87, 46 FR 40384, Aug. 7, 1981, as amended by T.D. ATF–290, 54 FR 53054, Dec. 27, 1989; ATF No. 1, 68 FR 13786, Mar. 20, 2003]

§ 555.58 Continuing partnerships.

Where, under the laws of the particular State, the partnership is not terminated on death or insolvency of a partner, but continues until the winding up of the partnership affairs is completed, and the surviving partner has the exclusive right to the control and possession of the partnership assets for the purpose of liquidation and settlement, the surviving partner may continue to conduct the business or operations under the license or permit of the partnership. If the surviving partner acquires the business or operations on completion of settlement of the partnership, he shall obtain a license or permit in his own name from the date of acquisition, as provided in §555.45. The rule set forth in this section will also apply where there is more than one surviving partner.

§ 555.59 Right of succession by certain persons.

(a) Certain persons other than the licensee or permittee may secure the right to carry on the same explosive materials business or operations at the same business premises for the remainder of the term of license or permit. These persons are:

(1) The surviving spouse or child, or executor, administrator, or other legal representative of a deceased licensee or permittee; and

(2) A receiver or trustee in bankruptcy, or an assignee for benefit of creditors.

(b) In order to secure the right of succession, the person or persons continuing the business or operations shall submit the license or permit and all copies furnished with the license or permit for endorsement of the succession to the Chief, Firearms and Explosives Licensing Center, within 30 days from the date on which the successor begins to carry on the business or operations.
[T.D. ATF–87, 46 FR 40384, Aug. 7, 1981, as amended by T.D. ATF–290, 54 FR 53054, Dec. 27, 1989]

§ 555.60 Certain continuances of business or operations.

A licensee or permittee who furnishes his license or permit to the Chief, Firearms and Explosives Licensing Center, for correction, amendment, or endorsement, as provided in this subpart, may continue his business or operations while awaiting its return.
[T.D. ATF–87, 46 FR 40384, Aug. 7, 1981, as amended by T.D. ATF–290, 54 FR 53054, Dec. 27, 1989]

§ 555.61 Discontinuance of business or operations.

Where an explosive materials business or operations is either discontinued or succeeded by a new owner, the owner of the business or operations discontinued or succeeded shall, within 30 days, furnish notification of the discontinuance or succession and submit his license or permit and any copies furnished with the license or permit to the Chief, Firearms and Explosives Licensing Center. (See also §555.128.)
[T.D. ATF–87, 46 FR 40384, Aug. 7, 1981, as amended by T.D. ATF–290, 54 FR 53054, Dec. 27, 1989]

§ 555.62 State or other law.

A license or permit issued under this part confers no right or privilege to conduct business or operations, including storage, contrary to State or other law. The holder of a license or permit issued under this part is not, by reason of the rights and privileges granted by that license or permit, immune from punishment for conducting an explosive materials business or operations in violation of the provisions of any State or other law. Similarly, compliance with the provisions of any State or other law affords no immunity under Federal law or regulations.

§ 555.63 Explosives magazine changes.

(a) General.

(1) The requirements of this section are applicable to magazines used for other than temporary (under 24 hours) storage of explosives.

(2) A magazine is considered suitable for the storage of explosives if the construction requirements of this part are met during the time explosives are stored in the magazine.

(3) A magazine is considered suitable for the storage of explosives if positioned in accordance with the applicable table of distances as specified in this part during the time explosives are stored in the magazine.

(4) For the purposes of this section, notification of the regional director (compliance) may be by telephone or in writing. However, if notification of the regional director (compliance) is in writing it must be at least three business days in advance of making changes in construction to an existing magazine or constructing a new magazine, and at least five business days in advance of using any reconstructed magazine or added magazine for the storage of explosives.

(b) Exception. Mobile or portable type 5 magazines are exempt from the requirements of paragraphs (c) and (d) of this section, but must otherwise be in compliance with paragraphs (a) (2) and (3) of this section during the time explosives are stored in such magazines.

(c) Changes in magazine construction. A licensee or permittee who intends to make changes in construction of an existing magazine shall notify the regional director (compliance) describing the proposed changes prior to making any changes. Unless otherwise advised by the regional director (compliance), changes in construction may commence after explosives are removed from the magazine. Explosives may not be stored in a reconstructed magazine before the regional director (compliance) has been notified in accordance with paragraph (a)(4) of this section that the changes have been completed.

(d) Magazines acquired or constructed after permit or license is issued. A licensee or permittee who intends to construct or acquire additional magazines shall notify the regional director (compliance) in accordance with paragraph (a)(4) of this section describing the additional magazines and the class and quantity of explosives to be stored in the magazine. Unless otherwise advised by the regional director (compliance), additional magazines may be constructed, or acquired magazines may be used for the storage of explosives. Explosives must not be stored in a magazine under construction. The regional director (compliance) must be notified that construction has been completed.

[T.D. ATF–87, 46 FR 40384, Aug. 7, 1981, as amended by T.D. ATF–400, 63 FR 45002, Aug. 24, 1998]

Subpart E—License and Permit Proceedings

§ 555.71 Opportunity for compliance.

Except in cases of willfulness or those in which the public interest requires otherwise, and the regional director (compliance) so alleges in the notice of denial of an application or revocation of a license or permit, no license or permit will be revoked or renewal application denied without first calling to the attention of the licensee or permittee the reasons for the contemplated action and affording him an opportunity to demonstrate or achieve compliance with all lawful requirements and to submit facts, arguments, or proposals of adjustment. The notice of contemplated action, ATF F 5400.12, will afford the licensee or permittee 15 days from the date of receipt of the notice to respond. If no response is received within the 15 days, or if after consideration of relevant matters presented by the licensee or permittee, the regional director (compliance) finds that the licensee or permittee is not likely to abide by the law and regulations, he will proceed as provided in §555.74.

[T.D. ATF–87, 46 FR 40384, Aug. 7 1981, as amended by T.D. ATF–446, 66 FR 16602, Mar. 27, 2001]

§ 555.72 Denial of initial application.

Whenever the regional director (compliance) has reason to believe that an applicant for an original license or permit is not eligible to receive a license or permit under the provisions of §555.49, he shall issue a notice of denial on ATF F 5400.11. The notice will set forth the matters of fact and law relied upon in determining that the application should be denied, and will afford the applicant 15 days from the date of receipt of the notice in which to request a hearing to review the denial. If no request for a hearing is filed within that time, a copy of the application, marked *"Disapproved"*, will be returned to the applicant.

§ 555.73 Hearing after initial application is denied.

If the applicant for an original license or permit desires a hearing, he shall file a request with the regional director (compliance) within 15 days after receipt of the notice of denial. The request should include a statement of the reasons for a hearing. On receipt of the request, the regional director (compliance) shall refer the matter to an administrative law judge who shall set a time and place (see §555.77) for a hearing and shall serve notice of the hearing upon the applicant and the regional director (compliance) at least 10 days in advance of the hearing date. The hearing will be conducted in accordance with the hearing procedures prescribed in part 71 of this chapter (see §555.82). Within a reasonable time after the conclusion of the hearing, and as expeditiously as possible, the administrative law judge shall render his recommended decision. He shall certify to the complete record of the proceedings before him and shall immediately forward the complete certified record, together with four copies of his recommended decision, to the regional director (compliance) for decision.

§ 555.74 Denial of renewal application or revocation of license or permit.

If following the opportunity for compliance under §555.71, or without opportunity for compliance under §555.71, as circumstances warrant, the regional director (compliance) finds that the licensee or permittee is not likely to comply with the law or regulations or is otherwise not eligible to continue operations authorized under his license or permit, the regional director (compliance) shall issue a notice of denial of the renewal application or revocation of the license or permit, ATF F 5400.11 or ATF F 5400.10, as appropriate. In either case, the notice will set forth the matters of fact constituting the violations specified, dates, places, and the sections of law and regulations violated. The notice will, in the case of revocation of a license or permit, specify the date on which the action is effective, which date will be on or after the date the notice is served on the licensee or permittee. The notice will also advise the licensee or permittee that he may, within 15 days after receipt of the notice, request a hearing and, if applicable, a stay of the effective date of the revocation of his license or permit.

§ 555.75 Hearing after denial of renewal application or revocation of license or permit.

If a licensee or permittee whose renewal application has been denied or whose license or permit has been revoked desires a hearing, he shall file a request for a hearing with the regional director (compliance). In the case of the revocation of a license or permit, he may include a request for a stay of the effective date of the revocation. On receipt of the request the regional director (compliance) shall advise the licensee or permittee whether the stay of the effective date of the revocation is granted. If the stay of the effective date of the revocation is granted, the regional director (compliance) shall refer the matter to an administrative law judge who shall set a time and place (see §555.77) for a hearing and shall serve notice of the hearing upon the licensee or permittee and the regional director (compliance) at least 10 days in advance of the hearing date. If the stay of the effective date of the revocation is denied, the licensee or permittee may request an immediate hearing. In this event, the regional director (compliance) shall immediately refer the matter to an administrative law judge who shall set a date and place for a hearing, which date shall be no later than 10 days from the date the licensee or permittee requested an immediate hearing. The hearing will be held in accordance with the applicable provisions of part 71 of this chapter. Within a reasonable time after the conclusion of the hearing, and as expeditiously as possible, the administrative law judge shall render his decision. He shall certify to the complete record of the proceeding before him and shall immediately forward the complete certified record, together with two copies of his decision, to the regional director (compliance), serve one copy of his decision on the licensee or permittee or his counsel, and transmit a copy to the attorney for the Government.

§ 555.76 Action by regional director (compliance).

(a) Initial application proceedings. If, upon receipt of the record and the recommended decision of the administrative law judge, the regional director (compliance) decides that the license or permit should be issued, the regional director (compliance) shall cause the application to be approved, briefly stating, for the record, his reasons. If he contemplates that the denial should stand, he shall serve a copy of the administrative law judge's recommended decision on the applicant, informing the applicant of his contemplated action and affording the applicant not more than 10 days in which to submit proposed findings and conclusions or exceptions to the recommended decision with supporting reasons. If the regional director (compliance), after consideration of the record of the hearing and of any proposed findings, conclusions, or exceptions filed with him by the applicant, approves the findings, conclusions and recommended decision of the administrative law judge, the regional director (compliance) shall cause the license or permit to be issued or disapproved the application accordingly. If he disapproves the findings, conclusions, and recommendation of the administrative law judge, in whole or in part, he shall by order make such findings and conclusions as in his opinion are warranted by the law and the facts in the record. Any decision of the regional director (compliance) ordering the disapproval of an initial application for a license or permit shall state the findings and conclusions upon which it is based, including his ruling upon each proposed finding, conclusion, and exception to the administrative law judge's recommended decision, together with a statement of his findings and conclusions, and reasons or basis for his findings and conclusions, upon all material issues of fact, law or discretion presented on the record. A signed duplicate original of the decision will be served upon the applicant and the original copy containing certificate of service will be placed in the official record of the proceedings. If the decision of the regional director (compliance) is in favor of the applicant, he shall issue the license or permit, to be effective on issuance.

(b) Renewal application and revocation proceedings. Upon receipt of the complete certified records of the hearing, the regional director (compliance) shall enter an order confirming the revocation of the license or permit, or disapproving the application, in accordance with the administrative law judge's findings and decision, unless he disagrees with the findings and decision. A signed duplicate original of the order, ATF F 5400.9, will be served upon the licensee or permittee and the original copy containing certificate of service will be placed in the official record of the proceedings. If the regional director (compliance) disagrees with the findings and decision of the administrative law judge, he shall file a petition with the Director for review of the findings and decision, as provided in §555.79. In either case, if the renewal application denial is sustained, a copy of the application marked "Disapproved" will be returned to the applicant. If the renewal application denial is reversed, a license or permit will be issued to become effective on expiration of the license or permit being renewed, or on the date of issuance, whichever is later. If the proceedings involve the revocation of a license or permit which expired before a decision is in favor of the licensee or permittee, the regional director (compliance) shall:

(1) If renewal application was timely filed and a stay of the effective date of the revocation was granted, cause to be issued a license or permit effective on the date of issuance;

(2) If renewal application was not timely filed but a stay of the effective date of the revocation had been granted, request that a renewal application be filed and, following that, cause to be issued a license or permit to be effective on issuance; or

(3) If a stay of the effective date of the revocation had not been granted, request that an application be filed as provided in §555.45, and process it in the same manner as for an application for an original license or permit.

(c) Curtailment of stay of revocation effective date. If, after approval of a request for a stay of the effective date of an order revoking a license or permit but before actions are completed under this subpart, the regional director (compliance) finds that it is contrary to the public interest for the licensee or permittee to continue the operations or activities covered by his license or permit, the regional director (compliance) may issue a notice of withdrawal of the approval, effective on the date of issuance. Notice of withdrawal will be served upon the licensee or permittee in the manner provided in §555.81.

[T.D. ATF–87, 46 FR 40384, Aug. 7, 1981, as amended by T.D. ATF–290, 54 FR 53054, Dec. 27, 1989]

§ 555.77 Designated place of hearing.

The designated place of hearing set as provided in §555.73 or §555.75, will be at the location convenient to the aggrieved party.

§ 555.78 Representation at a hearing.

An applicant, licensee, or permittee may be represented by an attorney, certified public accountant, or other person recognized to practice before the Bureau of Alcohol, Tobacco and Firearms as provided in 31 CFR Part 8, if he has otherwise complied with the applicable requirements of 26 CFR 601.521 through 601.527. The regional director (compliance) shall be represented in proceedings under §§555.73 and 555.75 by an attorney in the office of the chief counsel or regional counsel who is authorized to execute and file motions, briefs, and other papers in the proceedings, on behalf of the regional director (compliance), in his own name as *"Attorney for the Government"*.

[T.D. ATF–87, 46 FR 40384, Aug. 7, 1981, as amended by T.D. ATF–92, 46 FR 46916, Sept. 23, 1981]

§ 555.79 Appeal on petition to the Director.

An appeal to the Director is not required prior to filing an appeal with the U.S. Court of Appeals for judicial review. An appeal may be taken by the applicant, licensee, or permittee to the Director from a decision resulting from a hearing under §555.73 or §555.75. An appeal may also be taken by a regional director (compliance) from a decision resulting from a hearing under §555.75 as provided in §555.76(b). The appeal shall be taken by filing a petition for review on appeal with the Director within 15 days of the service of an administrative law judge's decision or an order. The petition will set forth facts tending to show (a) action of an arbitrary nature, (b) action without reasonable warrant in fact, or (c) action contrary to law and regulations. A copy of the petition will be filed with the regional director (compliance) or served on the applicant, licensee, or permittee, as the case may be. In the event of appeal, the regional director (compliance) shall immediately forward the complete original record, by certified mail, to the Director for his consideration, review, and disposition as provided in subpart I of part 71 of this chapter. When, on appeal, the Director affirms the initial decision of the regional director (compliance) or the administrative law judge, as the case may be, the initial decision will be final.

§ 555.80 Court review.

An applicant, licensee, or permittee may, within 60 days after receipt of the decision of the administrative law judge or the final order of the regional director (compliance) or the Director, file a petition for a judicial review of the decision, with the U.S. Court of Appeals for the district in which he resides or has his principal place of business. The Director, upon notification that a petition has been filed, shall have prepared a complete transcript of the record of the proceedings. The regional director (compliance) or the Director, as the case may be, shall certify to the correctness of the transcript of the record, forward one copy to the attorney for the Government in the review of the case, and file the original record of the proceedings with the original certificate in the U.S. Court of Appeals.

§ 555.81 Service on applicant, licensee, or permittee.

All notices and other formal documents required to be served on an applicant, licensee, or permittee under this subpart will be served by certified mail or by personal delivery. Where service is by personal delivery, the signed duplicate original copy of the formal document will be delivered to the applicant, licensee, or permittee, or, in the case of a corporation, partnership, or association, by delivering it to an officer, manager, or general agent, or to its attorney of record.

§ 555.82 Provisions of part 200 made applicable.

The provisions of subpart G of part 200 of this chapter, as well as those provisions of part 71 relative to failure to appear, withdrawal of an application or surrender of a permit, the conduct of hearings before an administrative law judge, and record of testimony, are hereby made applicable to application, license, and permit proceedings under this subpart to the extent that they are not contrary to or incompatible with this subpart.

§ 555.83 Operations by licensees or permittees after notice of denial or revocation.

In any case where a notice of revocation has been issued and a request for a stay of the effective date of the revocation has not been granted, the licensee or permittee shall not engage in the activities covered by the license or permit pending the outcome of proceedings under this subpart. In any case where notice of revocation has been issued but a stay of the effective date of the revocation has been granted, the licensee or permittee may continue to engage in the activities covered by his license or permit unless, or until, formally notified to the contrary: Provided, That in the event the license or permit would have expired before proceedings under this subpart are completed, timely renewal application must have been filed to continue the license or permit beyond its expiration date. In any case where a notice of denial of a renewal application has been issued, the licensee or permittee may continue to engage in the activities covered by the existing license or permit after the date of expiration of the license or permit until proceedings under this subpart are completed.

Subpart F—Conduct of Business or Operations

§ 555.101 Posting of license or user permit.

A license or user permit issued under this part, or a copy of a license or user permit, will be posted and available for inspection on the business premises at each place where explosive materials are manufactured, imported, or distributed.

[T.D. ATF–87, 46 FR 40384, Aug. 7, 1981. Redesignated by T.D. ATF–487, 68 FR 3748, Jan. 24, 2003. ATF No. 1, 68 FR 13786, Mar. 20, 2003]

§ 555.102 Authorized operations by permittees.

(a) In general. A permit issued under this part does not authorize the permittee to engage in the business of manufacturing, importing, or dealing in explosive materials. Accordingly, if a permittee's operations bring him within the definition of manufacturer, importer, or dealer under this part, he shall qualify for the appropriate license.

(b) Distributions of surplus stocks—

(1) Distributions of surplus stocks prior to May 24, 2003. Permittees are not authorized to engage in the business of sale or distribution of explosive materials. However, permittees may dispose of surplus stocks of explosive materials to other licensees or permittees in accordance with §555.103, and to nonlicensees or to nonpermittees in accordance with §555.105(a)(4).

(2) Distributions of surplus stocks on and after May 24, 2003. Permittees are not authorized to engage in the business of sale or distribution of explosive materials. However, permittees may dispose of surplus stocks of explosive materials to other licensees or permittees in accordance with §555.103 and §555.105.

[T.D. ATF–400, 63 FR 45002, Aug. 24, 1998, as amended by ATF No. 1, 68 FR 13787, Mar. 20, 2003]

§ 555.103 Transactions among licensees/permittees and transactions among licensees and holders of user permits.

(a) Transactions among licensees/permittees prior to May 24, 2003—

(1) General.

(i) A licensed importer, licensed manufacturer or licensed dealer selling or otherwise distributing explosive materials (or a permittee disposing of surplus stock to a licensee or another permittee) who has the certified information required by this section may sell or distribute explosive materials to a licensee or permittee for not more than 45 days following the expiration date of the distributee's license or permit, unless the distributor knows or has reason to believe that the distributee's authority to continue business or operations under this part has been terminated.

(ii) A licensed importer, licensed manufacturer or licensed dealer selling or otherwise distributing explosive materials (or a permittee disposing of surplus stock to another licensee or permittee) must verify the license or permit status of the distributee prior to the release of explosive materials ordered, as required by this section.

(iii) Licensees or permittees desiring to return explosive materials to a licensed manufacturer may do so without obtaining a certified copy of the manufacturer's license.

(iv) Where possession of explosive materials is transferred at the distributor's premises, the distributor must in all instances verify the identity of the person accepting possession on behalf of the distributee before relinquishing possession. Before the delivery at the distributor's premises of explosive materials to an employee of a licensee or permittee, or to an employee of a common or contract carrier transporting explosive materials to a licensee or permittee, the distributor delivering explosive materials must obtain an executed ATF F 5400.8, Explosives Delivery Record, from the employee before releasing the explosive materials. The ATF F 5400.8 must contain all of the information required on the form and required by this part.

Example 1. An ATF F 5400.8 is required when:
a. An employee of the purchaser takes possession at the distributor's premises.
b. An employee of a common or contract carrier hired by the purchaser takes possession at the distributor's premises.

Example 2. An ATF F 5400.8 is not required when:
a. An employee of the distributor takes possession of the explosives for the purpose of transport to the purchaser.
b. An employee of a common or contract carrier hired by the distributor takes possession of the explosives for the purpose of transport to the purchaser.

(2) License/permit verification of individuals.

(i) The distributee must furnish a certified copy (or, in the case of a user-limited, the original) of the license or permit. The certified copy need be furnished only once during the current term of the license or permit. Also, a licensee need not furnish certified copies of licenses to other licensed locations operated by such licensee.

(ii) The distributor may obtain any additional verification as the distributor deems necessary.

(3) License/permit verification of business organizations.

(i) A business organization may (in lieu of furnishing a certified copy of a license) furnish the distributor a certified list which contains the name, address, license number and date of license expiration of each licensed location. The certified list need be furnished only once during the current term of the license or permit. Also, a business organization need not furnish a certified list to other licensed locations operated by such business organization.

(ii) A business organization must, prior to ordering explosive materials, furnish the licensee or permittee a current certified list of the representatives or agents authorized to order explosive materials on behalf of the business organization showing the name, address, and date and place of birth of each representative or agent. A licensee or permittee may not distribute explosive materials to a business organization on the order of a person who does not appear on the certified list of representatives or agents and, if the person does appear on the certified list, the licensee or permittee must verify the identity of such person.

(4) Licensee/permittee certified statement.

(i) A licensee or permittee ordering explosive materials from another licensee or permittee must furnish a current, certified statement of the intended use of the explosive materials, e.g., resale, mining, quarrying, agriculture, construction, sport rocketry, road building, oil well drilling, seismographic research, to the distributor.

(ii) For individuals, the certified statement of intended use must specify the name, address, date and place of birth, and social security number of the distributee.

(iii) For business organizations, the certified statement of intended use must specify the taxpayer identification number, the identity and the principal and local places of **business.**

(iv) The licensee or permittee purchasing explosive materials must revise the furnished copy of the certified statement only when the information is no longer current.

(5) User-limited permit transactions. A user-limited permit issued under the provisions of this part is valid for only a single purchase transaction and is not renewable (see §555.51). Accordingly, at the time a user-limited permittee orders explosive materials, the licensed distributor must write on the front of the user-limited permit the transaction date, his signature, and the distributor's license number prior to returning the permit to the user-limited permittee.

(b) Transactions among licensees/permittees on and after May 24, 2003—

(1) General.

(i) A licensed importer, licensed manufacturer or licensed dealer selling or otherwise distributing explosive materials (or a holder of a user permit disposing of surplus stock to a licensee; a holder of a user permit; or a holder of a limited permit who is within the same State as the distributor) who has the certified information required by this section may sell or distribute explosive materials to a licensee or permittee for not more than 45 days following the expiration date of the distributee's license or permit, unless the distributor knows or has reason to believe that the distributee's authority to continue business or operations under this part has been terminated.

(ii) A licensed importer, licensed manufacturer or licensed dealer selling or otherwise distributing explosive materials (or a holder of a user permit disposing of surplus stock to another licensee or permittee) must verify the license or permit status of the distributee prior to the release of explosive materials ordered, as required by this section.

(iii) Licensees or permittees desiring to return explosive materials to a licensed manufacturer may do so without obtaining a certified copy of the manufacturer's license.

(2) Verification of license/user permit.

(i) Prior to or with the first order of explosive materials, the distributee must provide the distributor certified copy (or, in the case of a user-limited, the original) of the distributee's license or user permit. However, licensees or holders of user permits that are business organizations may (in lieu of a certified copy of a license or user permit) provide the distributor with a certified list that contains the name, address, license or user permit number, and date of the license or user permit expiration of each location.

(ii) The distributee must also provide the distributor with a current list of the names of persons authorized to accept delivery of explosive materials on behalf of the distributee. The distributee ordering explosive materials must keep the list current and provide updated lists to licensees and holders of user permits on a timely basis. A distributor may not transfer possession of explosive materials to any person whose name does not appear on the current list of names of persons authorized to accept delivery of explosive materials on behalf of the distributee. Except as provided in paragraph (b)(3) of this section, in all instances the distributor must verify the identity of the distributee, or the employee of the distributee accepting possession of explosive materials on behalf of the distributee, by examining an identification document (as defined in §555.11) before relinquishing possession.

(iii) A licensee or holder of a user permit ordering explosive materials from another licensee or permittee must provide to the distributor a current, certified statement of the intended use of the explosive materials, e.g., resale, mining, quarrying, agriculture, construction, sport rocketry, road building, oil well drilling, seismographic research, etc.

(A) For individuals, the certified statement of intended use must specify the name, address, date and place of birth, and social security number of the distributee.

(B) For business organizations, the certified statement of intended use must specify the taxpayer identification number, the identity and the principal and local places of business.

(C) The licensee or holder of a user permit purchasing explosive materials must revise the furnished copy of the certified statement only when the information is no longer current.

(3) Delivery of explosive materials by common or contract carrier. When a common or contract carrier will transport explosive materials from a distributor to a distributee who is a licensee or holder of a user permit, the distributor must take the following actions before relinquishing possession of the explosive materials:

(i) Verify the identity of the person accepting possession for the common or contract carrier by examining such person's valid, unexpired driver's license issued by any State, Canada, or Mexico; and

(ii) Record the name of the common or contract carrier (i.e., the name of the driver's employer) and the full name of the driver. This information must be maintained in the distributor's permanent records in accordance with §555.121.

(4) **User-limited permit transactions.** A user-limited permit issued under the provisions of this part is valid for only a single purchase transaction and is not renewable (see §555.51). Accordingly, at the time a user-limited permittee orders explosive materials, the licensed distributor must write on the front of the user-limited permit the transaction date, his signature, and the distributor's license number prior to returning the permit to the user-limited permittee.

(Approved by the Office of Management and Budget under control number 1140–0079)

[ATF No. 1, 68 FR 13787, Mar. 20, 2003, as amended by ATF No. 2, 68 FR 53512, Sept. 11, 2003]

§ 555.104 Certified copy of license or permit.

Except as provided in §555.49(a), each person issued a license or permit under this part shall be furnished together with his license or permit a copy for his certification. If a person desires an additional copy of his license or permit for certification and for use under §555.103, he shall:

(a) Make a reproduction of the copy of his license or permit and execute the certification on it;

(b) Make a reproduction of his license or permit, enter on the reproduction the statement: "I certify that this is a true copy of a (insert the word license or permit) issued to me to engage in the specified business or operations", and sign his name next to the statement; or

(c) Submit a request, in writing, for certified copies of his license or permit to the Chief, Firearms and Explosives Licensing Center. The request will show the name, trade name (if any), and address of the licensee or permittee and the number of copies of the license or permit desired. There is a fee of $1 for each copy of a license or permit issued by the Chief, Firearms and Explosives Licensing Center under this paragraph. Fee payment must accompany each request for additional copies of a license or permit. The fee must be paid by (1) cash, or (2) money order or check made payable to the Bureau of Alcohol, Tobacco and Firearms.

[T.D. ATF–87, 46 FR 40384, Aug. 7, 1981, as amended by T.D. ATF–290, 54 FR 53054, Dec. 27, 1989]

§ 555.105 Distributions to nonlicensees, nonpermittees, and limited permittees.

(a) Distributions to nonlicensees and nonpermittees prior to May 24, 2003.

(1) This section will apply in any case where distribution of explosive materials to the distributee is not otherwise prohibited by the Act or this part.

(2) Except as provided in paragraph (a)(3) of this section, a licensed importer, licensed manufacturer, or licensed dealer may distribute explosive materials to a nonlicensee or nonpermittee if the nonlicensee or nonpermittee is a resident of the same State in which the licensee's business premises are located, and the nonlicensee or nonpermittee furnishes to the licensee the explosives transaction record, ATF F 5400.4, required by §555.126. Disposition of ATF F 5400.4 will be made in accordance with §555.126.

(3) A licensed importer, licensed manufacturer, or licensed dealer may sell or distribute explosive materials to a resident of a State contiguous to the State in which the licensee's place of business is located if the purchaser's State of residence has enacted legislation, currently in force, specifically authorizing a resident of that State to purchase explosive materials in a contiguous State and the purchaser and the licensee have, prior to the distribution of the explosive materials, complied with all the requirements of paragraphs (a)(2), (a)(5), and (a)(6) of this section applicable to intrastate transactions occurring on the licensee's business premises.

(4) A permittee may dispose of surplus stocks of explosive materials to a nonlicensee or nonpermittee if the nonlicensee or nonpermittee is a resident of the same State in which the permittee's business premises or operations are located, or is a resident of a State contiguous to the State in which the permittee's place of business or operations are located, and if the requirements of paragraphs (a)(2), (a)(3), (a)(5), and (a)(6) of this section are fully met.

(5) A licensed importer, licensed manufacturer, or licensed dealer selling or otherwise distributing explosive materials to a business entity must verify the identity of the representative or agent of the business entity who is authorized to order explosive materials on behalf of the business entity. Each business entity ordering explosive materials must furnish the distributing licensee prior to or with the first order of explosive materials a current certified list of the names of representatives or agents authorized to order explosive materials on behalf of the business entity. The business entity ordering explosive materials is responsible for keeping the certified list current. A licensee may not distribute explosive materials to a business entity on the order of a person whose name does not appear on the certified list.

(6) Where the possession of explosive materials is transferred at the distributor's premises, the distributor must in all instances verify the identity of the person accepting possession on behalf of the distributee before relinquishing possession. Before the delivery at the distributor's premises of explosive materials to an employee of a nonlicensee or nonpermittee, or to an employee of a common or contract carrier transporting explosive materials to a nonlicensee or nonpermittee, the distributor

delivering explosive materials must obtain an executed ATF F 5400.8 from the employee before releasing the explosive materials. The ATF F 5400.8 must contain all of the information required on the form and by this part. (See examples in **§555.103(a)).**

(7) A licensee or permittee disposing of surplus stock may sell or distribute commercially manufactured black powder in quantities of 50 pounds or less to a nonlicensee or nonpermittee if the black powder is intended to be used solely for sporting, recreational, or cultural purposes in antique firearms as defined in 18 U.S.C. 921(a)(16), or in antique devices as exempted from the term "destructive device" in 18 U.S.C. 921(a)(4).

(b) Distributions to holders of limited permits on and after May 24, 2003.

(1) This section will apply in any case where distribution of explosive materials to the distributee is not otherwise prohibited by the Act or this part.

(2) A licensed importer, licensed manufacturer or a licensed dealer may distribute explosive materials to a holder of a limited permit if such permittee is a resident of the same State in which the licensee's business premises are located, the holder of the limited permit presents in person or by mail ATF Form 5400.4, Limited Permittee Transaction Report (LPTR), and the licensee completes Form 5400.4 in accordance with §555.126(b). In no event will a licensee distribute explosive materials to a holder of a limited permit unless the holder presents a Form 5400.4 with an original unaltered and unexpired Intrastate Purchase of Explosives Coupon (IPEC), ATF Form 5400.30, affixed. The coupon must bear the name, address, permit number, and the coupon number of the limited permittee seeking distribution of the explosives.

(3) A holder of a limited permit is authorized to receive explosive materials from a licensee or permittee whose premises are located in the same State of residence in which the premises of the holder of the limited permit are located on no more than 6 separate occasions during the one-year period of the permit. For purposes of this section, the term "6 separate occasions" means six deliveries of explosive materials. Each delivery must—

(i) Relate to a single purchase transaction made on one ATF F 5400.4;

(ii) Be referenced on one commercial invoice or purchase order; and

(iii) Be delivered to the holder of the limited permit in one shipment delivered at the same time.

(4) A holder of a user permit may dispose of surplus stocks of explosive materials to a licensee or holder of a user permit, or a holder of a limited permit who is a resident of the same State in which the premises of the holder of the user permit are located. A holder of a limited permit may dispose of surplus stocks of explosive materials to another holder of a limited permit who is a resident of the same State in which the premises of the distributor are located, if the transaction complies with the requirements of paragraph (b)(2) of this section and §555.126(b). A holder of a limited permit may also dispose of surplus stocks of explosive materials to a licensee or holder of a user permit if the disposition occurs in the State of residence of the holder of the limited permit. (See §555.103.)

(5) Each holder of a limited permit ordering explosive materials must furnish the distributing licensee prior to or with the first order of the explosive materials a current list of the names of employees authorized to accept delivery of explosive materials on behalf of the limited permittee. The distributee ordering explosive materials must keep the list current and provide updated lists to licensees and holders of user permits on a timely basis. A licensed importer, licensed manufacturer, licensed dealer, or permittee, selling or otherwise distributing explosive materials to a holder of a limited permit must, prior to delivering the explosive materials, obtain from the limited permittee a current list of persons who are authorized to accept deliveries of explosive materials on behalf of the limited permittee. A licensee or permittee may not deliver explosive materials to a person whose name does not appear on the list.

(6)(i) Delivery at the distributor's premises. Where possession of explosive materials is transferred directly to the distributee at the distributor's premises, the distributor must obtain an executed Form 5400.4 in accordance with §555.126(b) and must in all instances verify the identity of the person accepting possession on behalf of the distributee by examining an identification document (as defined in §555.11) before relinquishing possession.

(ii) Delivery by distributor. Where possession of explosive materials is transferred by the distributor to the distributee away from the distributor's premises, the distributor must obtain an executed Form 5400.4 in accordance with §555.126(b) and must in all instances verify the identity of the person accepting possession on behalf of the distributee by examining an identification document (as defined in §555.11) before relinquishing possession.

(iii) Delivery by common or contract carrier hired by the distributor. Where a common or contract carrier hired by the distributor will transport explosive materials from the distributor to a holder of a limited permit:

(A) The limited permittee must, prior to delivery of the explosive materials, complete the appropriate section on Form 5400.4, affix to the Form 5400.4 one of the six IPECs he has been issued, and provide the form to the distributor in person or by mail.

(B) The distributor must, before relinquishing possession of the explosive materials to the common or contract carrier:

(1) Verify the identity of the person accepting possession for the common or contract carrier by examining such person's valid, unexpired driver's license issued by any State, Canada, or Mexico; and

(2) Record the name of the common or contract carrier (i.e., the name of the driver's employer) and the full name of the driver. This information must be maintained in the distributor's permanent records in accordance with §555.121.

(C) At the time of delivery of the explosive materials, the common or contract carrier, as agent for the distributor, must verify the identity of the person accepting delivery on behalf of the distributee, note the type and number of the identification document (as defined in §555.11) and provide this information to the distributor. The distributor must enter this information in the appropriate section on Form 5400.4.

(iv) Delivery by common or contract carrier hired by the distributee. Where a common or contract carrier hired by the distributee will transport explosive materials from the distributor to a holder of a limited permit:

(A) The limited permittee must, prior to delivery of the explosive materials, complete the appropriate section on Form 5400.4, affix to the Form 5400.4 one of the six IPECs he has been issued, and provide the form to the distributor in person or by mail.

(B) Before the delivery at the distributor's premises to the common or contract carrier who will transport explosive materials to the holder of a limited permit, the distributor must:

(1) Verify the identity of the person accepting possession for the common or contract carrier by examining such person's valid, unexpired driver's license issued by any State, Canada, or Mexico; and

(2) Record the name of the common or contract carrier (i.e., the name of the driver's employer) and the full name of the driver. This information must be maintained in the distributor's permanent records in accordance with §555.121.

(7) A licensee or permittee disposing of surplus stock may sell or distribute commercially manufactured black powder in quantities of 50 pounds or less to a holder of a limited permit, nonlicensee, or nonpermittee if the black powder is intended to be used solely for sporting, recreational, or cultural purposes in antique firearms as defined in 18 U.S.C. 921(a)(16), or in antique devices as exempted from the term "destructive device" in 18 U.S.C. 921(a)(4).

(Approved by the Office of Management and Budget under control number 1140–0075)

[ATF No. 1, 68 FR 13788, Mar. 20, 2003, as amended by ATF No. 2, 68 FR 53513, Sept. 11, 2003]

§ 555.106 Certain prohibited distributions.

(a) A licensee or permittee may not distribute explosive materials to any person except—

(1) A licensee;

(2) A holder of a user permit; or

(3) A holder of a limited permit who is a resident of the State where distribution is made and in which the premises of the transferor are located.

(b) A licensee shall not distribute any explosive materials to any person:

(1) Who the licensee knows is less than 21 years of age;

(2) In any State where the purchase, possession, or use by a person of explosive materials would be in violation of any State law or any published ordinance applicable at the place of distribution;

(3) Who the licensee has reason to believe intends to transport the explosive materials into a State where the purchase, possession, or use of explosive materials is prohibited or which does not permit its residents to transport or ship explosive materials into the State or to receive explosive materials in the State; or

(4) Who the licensee has reasonable cause to believe intends to use the explosive materials for other than a lawful purpose.

(c) A licensee shall not distribute any explosive materials to any person knowing or having reason to believe that the person:

(1) Is, except as provided under §555.142 (d) and (e), under indictment or information for, or was convicted in any court of, a crime punishable by imprisonment for a term exceeding 1 year;

(2) Is a fugitive from justice;

(3) Is an unlawful user of marijuana, or any depressant or stimulant drug, or narcotic drug (as these terms are defined in the Controlled Substances Act, 21 U.S.C. 802);

(4) Was adjudicated as a mental defective or was committed to a mental institution;

(5) Is an alien, other than an alien who—

(i) Is lawfully admitted for permanent residence (as that term is defined in section 101(a)(20) of the Immigration and Nationality Act (8 U.S.C. 1101));

(ii) Is in lawful nonimmigrant status, is a refugee admitted under section 207 of the Immigration and Nationality Act (8 U.S.C. 1157), or is in asylum status under section 208 of the Immigration and Nationality Act (8 U.S.C. 1158), and—

(A) Is a foreign law enforcement officer of a friendly foreign government, as determined by the Attorney General in consultation with the Secretary of State, entering the United States on official law enforcement business, and the shipping, transporting, possession, or receipt of explosive materials is in furtherance of this official law enforcement business;

(B) Is a person having the power to direct or cause the direction of the management and policies of a corporation, partnership, or association licensed pursuant to section 843(a), and the shipping, transporting, possession, or receipt of explosive materials is in furtherance of such power;

(C) Is a member of a North Atlantic Treaty Organization (NATO) or other friendly foreign military force, as determined by the Attorney General in consultation with the Secretary of Defense, (whether or not admitted in a nonimmigrant status) who is present in the United States under military orders for training or other military purpose authorized by the United States, and the shipping, transporting, possession, or receipt of explosive materials is in furtherance of the military purpose; or

(D) Is lawfully present in the United States in cooperation with the Director of Central Intelligence, and the shipment, transportation, receipt, or possession of the explosive materials is in furtherance of such cooperation;

(6) Has been discharged from the armed forces under dishonorable conditions; or

(7) Having been a citizen of the United States, has renounced citizenship.

(d) The provisions of this section do not apply to the purchase of commercially manufactured black powder in quantities not to exceed 50 pounds, intended to be used solely for sporting, recreational, or cultural purposes in antique firearms or in antique devices, if the requirements of §555.105(a)(7) or (b)(7) are fully met.

[T.D. ATF–87, 46 FR 40384, Aug. 7, 1981. Redesignated by T.D. ATF–487, 68 FR 3748, Jan. 24, 2003. ATF No. 1, 68 FR 13790, Mar. 20, 2003]

§ 555.107 Record of transactions.

Each licensee and permittee shall keep records of explosive materials as required by subpart G of this part.

§ 555.108 Importation.

(a) Explosive materials imported or brought into the United States by a licensed importer or holder of a user permit may be released from customs custody to the licensed importer or holder of a user permit upon proof of his status as a licensed importer or holder of a user permit. Proof of status must be made by the licensed importer or holder of a user permit furnishing to the customs officer a certified copy of his license or permit (see §555.103).

(b) A nonlicensee or nonpermittee may import or bring into the United States commercially manufactured black powder in quantities not to exceed 50 pounds. Upon submitting to the customs officer completed ATF F 5400.3, certifying that the black powder is intended to be used solely for sporting, recreational, or cultural purposes in antique firearms or in antique devices, black powder may be released from customs custody. The disposition of the executed ATF F 5400.3 will be in accordance with the instructions on the form.

(c) The provisions of this section are in addition to, and are not in lieu of, any applicable requirement under 27 CFR Part 447.

(d) For additional requirements relating to the importation of plastic explosives into the United States on or after April 24, 1997, see §555.183.

(e) For requirements relating to the marking of imported explosive materials, see §555.109.

[T.D. ATF–87, 46 FR 40384, Aug. 7, 1981, as amended by T.D. ATF–387, 62 FR 8376, Feb. 25, 1997; ATF No. 1, 68 FR 13790, Mar. 20, 2003; ATF 5F, 70 FR 30633, May 27, 2005]

§ 555.109 Identification of explosive materials.

(a) General. Explosive materials, whether manufactured in the United States or imported, must contain certain marks of identification.

(b) Required marks

(1) Licensed manufacturers. Licensed manufacturers who manufacture explosive materials for sale or distribution must place the following marks of identification on explosive materials at the time of manufacture:

(i) The name of the manufacturer; and

(ii) The location, date, and shift of manufacture. Where a manufacturer operates his plant for only one shift during the day, he does not need to show the shift of manufacture.

(2) Licensed importers.

(i) Licensed importers who import explosive materials for sale or distribution must place the following marks of identification on the explosive materials they import:

(A) The name and address (city and state) of the importer; and

(B) The location (city and country) where the explosive materials were manufactured, date, and shift of manufacture. Where the foreign manufacturer operates his plant for only one shift during the day, he does not need to show the shift of manufacture.

(ii) Licensed importers must place the required marks on all explosive materials imported prior to distribution or shipment for use, and in no event later than 15 days after the date of release from Customs custody.

(c) General requirements.

(1) The required marks prescribed in this section must be permanent and legible.

(2) The required marks prescribed in this section must be in the English language, using Roman letters and Arabic numerals.

(3) Licensed manufacturers and licensed importers must place the required marks on each cartridge, bag, or other immediate container of explosive materials that they manufacture or import, as well as on any outside container used for the packaging of such explosive materials.

(4) Licensed manufacturers and licensed importers may use any method, or combination of methods, to affix the required marks to the immediate container of explosive materials, or outside containers used for the packaging thereof, provided the identifying marks are legible, permanent, show all the required information, and are not rendered unreadable by extended periods of storage.

(5) If licensed manufacturers or licensed importers desire to use a coding system and omit printed markings on the container that show all the required information specified in paragraphs (b)(1) and (2) of this section, they must file with ATF a letterhead application displaying the coding that they plan to use and explaining the manner of its application. The Director must approve the application before the proposed coding can be used.

(d) Exceptions —

(1) Blasting caps. Licensed manufacturers or licensed importers are only required to place the identification marks prescribed in this section on the containers used for the packaging of blasting caps.

(2) Alternate means of identification. The Director may authorize other means of identifying explosive materials, including fireworks, upon receipt of a letter application from the licensed manufacturer or licensed importer showing that such other identification is reasonable and will not hinder the effective administration of this part.

(Paragraph (b)(2) approved by the Office of Management and Budget under control number 1140–0055)

[ATF 5F, 70 FR 30633, May 27, 2005]

§ 555.110 Furnishing of samples (Effective on and after January 24, 2003).

(a) In general. Licensed manufacturers and licensed importers and persons who manufacture or import explosive materials or ammonium nitrate must, when required by letter issued by the Director, furnish—

(1) Samples of such explosive materials or ammonium nitrate;

(2) Information on chemical composition of those products; and

(3) Any other information that the Director determines is relevant to the identification of the explosive materials or to identification of the ammonium nitrate.

(b) Reimbursement. The Director will reimburse the fair market value of samples furnished pursuant to paragraph (a) of this section, as well as reasonable costs of shipment.

(Approved by the Office of Management and Budget under control number 1140–0073)

[ATF No. 1, 68 FR 13790, Mar. 20, 2003]

Subpart G—Records and Reports

§ 555.121 General.

(a) **(1)** Licensees and permittees shall keep records pertaining to explosive materials in permanent form (i.e., commercial invoices, record books) and in the manner required in this subpart.

(2) Licensees and permittees shall keep records required by this part on the business premises for five years from the date a transaction occurs or until discontinuance of business or operations by the licensee or permittee. (See also §555.128 for discontinuance of business or operations.)

(b) ATF officers may enter the premises of any licensee or holder of a user permit for the purpose of examining or inspecting any record or document required by or obtained under this part (see §555.24). Section 843(f) of the Act requires licensees and holders of user permits to make all required records available for examination or inspection at all reasonable times. Section 843(f) of the Act also requires licensees and permittees (including holders of limited permits) to submit all reports and information relating to all required records and their contents, as the regulations in this part prescribe.

(c) Each licensee and permittee shall maintain all records of importation, production, shipment, receipt, sale, or other disposition, whether temporary or permanent, of explosive materials as the regulations in this part prescribe. Sections 842(f) and 842(g) of the Act make it unlawful for any licensee or permittee knowingly to make any false entry in, or fail to make entry in, any record required to be kept under the Act and the regulations in this part.

(Approved by the Office of Management and Budget under control number 1512–0373)

[T.D. ATF–87, 46 FR 40384, Aug. 7, 1981, as amended by T.D. ATF–172, 49 FR 14941, Apr. 16, 1984; ATF No. 1, 68 FR 13790, Mar. 20, 2003]

§ 555.122 Records maintained by licensed importers.

(a) Each licensed importer shall take true and accurate physical inventories which will include all explosive materials on hand required to be accounted for in the records kept under this part. The licensed importer shall take a special inventory

(1) At the time of commencing business, which is the effective date of the license issued upon original qualification under this part;

(2) At the time of changing the location of his business to another region;

(3) At the time of discontinuing business; and

(4) At any time the regional director (compliance) may in writing require. Each special inventory is to be prepared in duplicate, the original of which is submitted to the regional director (compliance), and the duplicate retained by the licensed importer. If a special inventory specified by paragraphs (a) (1) through (4) of this section has not been taken during the calendar year, at least one physical inventory will be taken. However, the record of the yearly inventory, other than a special inventory required by paragraphs (a) (1) through (4) of this section, will remain on file for inspection instead of being sent to the regional director (compliance). (See also §555.127.)

(b) Each licensed importer shall, not later than the close of the next business day following the date of importation or other acquisition of explosive materials, enter the following information in a separate record:

(1) Date of importation or other acquisition.

(2) Name or brand name of manufacturer and country of manufacture.

(3) Manufacturer's marks of identification.

(4) Quantity (applicable quantity units, such as pounds of explosives, number of detonators, number of display fireworks, etc.).

(5) Description (dynamite (dyn), blasting agents (ba), detonators (det), display fireworks (df), etc.) and size (length and diameter or diameter only of display fireworks).

(c) Each licensed importer shall, not later than the close of the next business day following the date of distribution of any explosive materials to another licensee or a permittee, enter in a separate record the following information:

(1) Date of disposition.

(2) Name or brand name of manufacturer and country of manufacture.

(3) Manufacturer's marks of identification.

(4) Quantity (applicable quantity units, such as pounds of explosives, number of detonators, number of display fireworks, etc.).

(5) Description (dynamite (dyn), blasting agents (ba), detonators (det), display fireworks (df), etc.) and size (length and diameter or diameter only of display fireworks).

(6) License or permit number of licensee or permittee to whom the explosive materials are distributed.

(d) The regional director (compliance) may authorize alternate records to be maintained by a licensed importer to record his distribution of explosive materials when it is shown by the licensed importer that alternate records will accurately and readily disclose the information required by paragraph (c) of this section. A licensed importer who proposes to use alternate records shall submit a letter application to the regional director (compliance) and shall describe the proposed alternate records and the need for them. Alternate records are not to be employed by the licensed importer until approval is received from the regional director (compliance).

(e) Each licensed importer shall maintain separate records of the sales or other distribution made of explosive materials to nonlicensees or nonpermittees. These records are maintained as prescribed by §555.126.

(Approved by the Office of Management and Budget under control number 1512–0373)

[T.D. ATF–87, 46 FR 40384, Aug. 7, 1981, as amended by T.D. ATF–172, 49 FR 14941, Apr. 16, 1984; T.D. ATF–293, 55 FR 3721, Feb. 5, 1990; T.D. ATF–400, 63 FR 45003, Aug. 24, 1998]

§ 555.123 Records maintained by licensed manufacturers.

(a) Each licensed manufacturer shall take true and accurate physical inventories which will include all explosive materials on hand required to be accounted for in the records kept under this part. The licensed manufacturer shall take a special inventory

(1) At the time of commencing business, which is the effective date of the license issued upon original qualification under this part;

(2) At the time of changing the location of his premises to another region;

(3) At the time of discontinuing business; and

(4) At any other time the regional director (compliance) may in writing require. Each special inventory is to be prepared in duplicate, the original of which is submitted to the regional director (compliance), and the duplicate retained by the licensed manufacturer. If a special inventory required by paragraphs (a) (1) through (4) of this section has not been taken during the calendar year, at least one physical inventory will be taken. However, the record of the yearly inventory, other than a special inventory required by paragraphs (a) (1) through (4) of this section, will remain on file for inspection instead of being sent to the regional director (compliance). (See also §555.127.)

(b) Each licensed manufacturer shall not later than the close of the next business day following the date of manufacture or other acquisition of explosive materials, enter the following information in a separate record:

(1) Date of manufacture or other acquisition.

(2) Manufacturer's marks of identification.

(3) Quantity (applicable quantity units, such as pounds of explosives, number of detonators, number of display fireworks, etc.).

(4) Name, brand name or description (dynamite (dyn), blasting agents (ba), detonators (det), display fireworks (df), etc.) and size (length and diameter or diameter only of display fireworks).

(c) Each licensed manufacturer shall, not later than the close of the next business day following the date of distribution of any explosive materials to another licensee or a permittee, enter in a separate record the following information:

(1) Date of disposition.

(2) Name or brand name of manufacturer or name of importer, as applicable, if acquired other than by his own manufacture.

(3) Manufacturer's marks of identification.

(4) Quantity (applicable quantity units, such as pounds of explosives, number of detonators, number of display fireworks, etc.).

(5) Description (dynamite (dyn), blasting agents (ba), detonators (det), display fireworks (df), etc.) and size (length and diameter or diameter only of display fireworks).

(6) License or permit number of licensee or permittee to whom the explosive materials are distributed.

(d) Each licensed manufacturer who manufactures explosive materials for his own use shall, not later than the close of the next business day following the date of use, enter in a separate record the following information:

(1) Date of use.

(2) Quantity (applicable quantity units, such as pounds of explosives, number of detonators, number of special fireworks, etc.).

(3) Description (dynamite (dyn), blasting agents (ba), detonators (det), display fireworks (df), etc.) and size (length and diameter or diameter only of display fireworks).

Exception: A licensed manufacturer is exempt from the recordkeeping requirements of this subsection if the explosive materials are manufactured for his own use and used within a 24 hour period at the same site.

(e) The regional director (compliance) may authorize alternate records to be maintained by a licensed manufacturer to record his distribution or use of explosive materials when it is shown by the licensed manufacturer that alternate records will accurately and readily disclose the information required by paragraph (c) of this section. A licensed manufacturer who proposes to use alternate records shall submit a letter application to the regional director (compliance) and shall describe the proposed alternate records and the need for them. Alternate records are not to be employed by the licensed manufacturer until approval is received from the regional director (compliance).

(f) Each licensed manufacturer shall maintain separate records of the sales or other distribution made of explosive materials to nonlicensees or nonpermittees. These records are maintained as prescribed by §555.126.

(Approved by the Office of Management and Budget under control number 1512–0373)

[T.D. ATF–87, 46 FR 40384, Aug. 7, 1981, as amended by T.D. ATF–172, 49 FR 14941, Apr. 16, 1984; T.D. ATF–293, 55 FR 3721, Feb. 5, 1990; T.D. ATF–400, 63 FR 45003, Aug. 24, 1998]

§ 555.124 Records maintained by licensed dealers.

(a) Each licensed dealer shall take true and accurate physical inventories which will include all explosive materials on hand required to be accounted for in the records kept under this part. The licensed dealer shall take a special inventory

(1) At the time of commencing business, which is the effective date of the license issued upon original qualification under this part;

(2) At the time of changing the location of his premises to another region;

(3) At the time of discontinuing business; and

(4) At any other time the regional director (compliance) may in writing require. Each special inventory is to be prepared in duplicate, the original of which is submitted to the regional director (compliance), and the duplicate retained by the licensed dealer. If a special inventory required by paragraphs (a)(1) through (4) of this section has not been taken during the calendar year, at least one physical inventory will be taken. However, the record of the yearly inventory, other than a special inventory required by paragraphs (a)(1) through (4) of this section, will remain on file for inspection instead of being sent to the regional director (compliance). (See also §555.127.)

(b) Each licensed dealer shall, not later than the close of the next business day following the date of purchase or other acquisition of explosive materials (except as provided in paragraph (d) of this section), enter the following information in a separate record:

(1) Date of acquisition.

(2) Name or brand name of manufacturer and name of importer (if any).

(3) Manufacturer's marks of identification.

(4) Quantity (applicable quantity units, such as pounds of explosives, number of detonators, number of display fireworks, etc.).

(5) Description (dynamite (dyn), blasting agents (ba), detonators (det), display fireworks (df), etc.) and size (length and diameter or diameter only of display fireworks).

(6) Name, address, and license or permit number of the person from whom the explosive materials are received.

(c) Each licensed dealer shall, not later than the close of the next business day following the date of use (if the explosives are used by the dealer) or the date of distribution of any explosive materials to another licensee or a permittee (except as provided in paragraph (d) of this section), enter in a separate record the following information:

(1) Date of disposition.

(2) Name or brand name of manufacturer and name of importer (if any).

(3) Manufacturer's marks of identification.

(4) Quantity (applicable quantity units, such as pounds of explosives, number of detonators, number of display fireworks, etc.).

(5) Description (dynamite (dyn), blasting agents (ba), detonators (det), display fireworks (df), etc.) and size (length and diameter or diameter only of display fireworks).

(6) License or permit number of licensee or permittee to whom the explosive materials are distributed.

(d) When a commercial record is kept by a licensed dealer showing the purchase or other acquisition information required for the permanent record prescribed by paragraph (b) of this section, or showing the distribution information required for the permanent record prescribed by paragraph (c) of this section, the licensed dealer acquiring or distributing the explosive materials may, for a period not exceeding seven days following the date of acquisition of distribution of the explosive materials, delay making the required entry into the permanent record of acquisition or distribution. However, until the required entry of acquisition or disposition is made in the permanent record, the commercial record must be (1) kept by the licensed dealer separate from other commercial documents kept by the licensee, and (2) readily available for inspection on the licensed premises.

(e) The regional director (compliance) may authorize alternate records to be maintained by a licensed dealer to record his acquisition or disposition of explosive materials, when it is shown by the licensed dealer that alternate records will accurately and readily disclose the required information. A licensed dealer who proposes to use alternate records shall submit a letter application to the regional director (compliance) and shall describe the proposed alternate records and the need for them. Alternate records are not to be employed by the licensed dealer until approval is received from the regional director (compliance).

(f) Each licensed dealer shall maintain separate records of the sales or other distribution made of explosive materials to nonlicensees or nonpermittees. These records are maintained as prescribed by §555.126.

(Approved by the Office of Management and Budget under control number 1512–0373)

[T.D. ATF–87, 46 FR 40384, Aug. 7, 1981, as amended by T.D. ATF–172, 49 FR 14941, Apr. 16, 1984; T.D. ATF–293, 55 FR 3721, Feb. 5, 1990; T.D. ATF–400, 63 FR 45003, Aug. 24, 1998]

§ 555.125 Records maintained by permittees.

(a) Records maintained by permittees prior to May 24, 2003.

(1) Each permittee must take true and accurate physical inventories that will include all explosive materials on hand required to be accounted for in the records kept under this part. The permittee must take a special inventory—

(i) At the time of commencing business, which is the effective date of the permit issued upon original qualification under this part;

(ii) At the time of changing the location of his premises to another region;

(iii) At the time of discontinuing business; and

(iv) At any other time the regional director (compliance) may in writing require. Each special inventory is to be prepared in duplicate, the original of which is submitted to the regional director (compliance) and the duplicate retained by the permittee. If a special inventory required by paragraphs (a)(1)(i) through (iv) of this section has not been taken during the calendar year, a permittee is required to take at least one physical inventory. However, the record of the yearly inventory, other than a special inventory required by paragraphs (a)(1)(i) through (iv) of this section, will remain on file for inspection instead of being sent to the regional director (compliance). (See also §555.127.)

(2) Each permittee must, not later than the close of the next business day following the date of acquisition of explosive materials, enter the following information in a separate record:

(i) Date of acquisition;

(ii) Name or brand name of manufacturer;

(iii) Manufacturer's marks of identification;

(iv) Quantity (applicable quantity units, such as pounds of explosives, number of detonators, number of display fireworks, etc.);

(v) Description (dynamite (dyn), blasting agents (ba), detonators (det), display fireworks (df), etc., and size (length and diameter or diameter only of display fireworks)); and

(vi) Name, address, and license number of the persons from whom the explosive materials are received.

(3) Each permittee must, not later than the close of the next business day following the date of disposition of surplus explosive materials to another permittee or a licensee, enter in a separate record the information prescribed in §555.124(c).

(4) Each permittee must maintain separate records of disposition of surplus stocks of explosive materials to nonlicensees or nonpermittees as prescribed in §555.126.

(5) The regional director (compliance) may authorize alternate records to be maintained by a permittee to record his acquisition of explosive materials, when it is shown by the permittee that alternate records will accurately and readily disclose the required information. A permittee who proposes to use alternate records must submit a letter application to the regional director (compliance) and must describe the proposed alternate records and the need for them. Alternate records are not to be employed by the permittee until approval is received from the regional director (compliance).

(b) Records maintained by permittees on and after May 24, 2003.

(1) Each holder of a user permit must take true and accurate physical inventories that will include all explosive materials on hand required to be accounted for in the records kept under this part. The permittee must take a special inventory—

(i) At the time of commencing business, which is the effective date of the permit issued upon original qualification under this part;

(ii) At the time of changing the location of his premises;

(iii) At the time of discontinuing business; and

(iv) At any other time the regional director (compliance) may in writing require. Each special inventory is to be prepared in duplicate, the original of which is submitted to the regional director (compliance) and the duplicate retained by the permittee. If a special inventory required by paragraphs (b)(1)(i) through (iv) of this section has not been taken during the calendar year, a permittee is required to take at least one physical inventory.

The record of the yearly inventory, other than a special inventory required by paragraphs (b)(1)(i) through (iv) of this section, will remain on file for inspection instead of being sent to the regional director (compliance). (See also §555.127.)

(2) Each holder of a limited permit must take true and accurate physical inventories, at least annually, that will include all explosive materials on hand required to be accounted for in the records kept under this part.

(3) Each holder of a user permit or a limited permit must, not later than the close of the next business day following the date of acquisition of explosive materials, enter the following information in a separate record:

(i) Date of acquisition;

(ii) Name or brand name of manufacturer;

(iii) Manufacturer's marks of identification;

(iv) Quantity (applicable quantity units, such as pounds of explosives, number of detonators, number of display fireworks, etc.);

(v) Description (dynamite (dyn), blasting agents (ba), detonators (det), display fireworks (df), etc., and size (length and diameter or diameter only of display fireworks)); and

(vi) Name, address, and license number of the persons from whom the explosive materials are received.

(4) Each holder of a user permit or a limited permit must, not later than the close of the next business day following the date of disposition of surplus explosive materials to another permittee or a licensee, enter in a separate record the information prescribed in §555.124(c).

(5) When a record book is used as a permittee's permanent record the permittee may delay entry of the required information for a period not to exceed seven days if the commercial record contains all of the required information prescribed by paragraphs (b)(3) and (b)(4) of this section. However, the commercial record may be used instead of a record book as a permanent record provided that the record contains all of the required information prescribed by paragraphs (b)(3) and (b)(4) of this section.

(6) Each holder of a user permit or a limited permit must maintain separate records of disposition of surplus stocks of explosive materials to holders of a limited permit as prescribed in §555.126.

(7) The regional director (compliance) may authorize alternate records to be maintained by a holder of a user permit or a limited permit to record his acquisition of explosive materials, when it is shown by the permittee that alternate records will accurately and readily disclose the required information. A permittee who proposes to use alternate records must submit a letter application to the regional director (compliance) and must describe the proposed alternate records and the need for them. Alternate records are not to be employed by the permittee until approval is received from the regional director (compliance).

(Approved by the Office of Management and Budget under control number 1140–0030)

[ATF No. 1, 68 FR 13790, Mar. 20, 2003]

§ 555.126 Explosives transaction record for distribution of explosive materials prior to May 24, 2003 and Limited Permittee Transaction Report for distribution of explosive materials on and after May 24, 2003.

(a) Explosives transaction record for distribution of explosive materials prior to May 24, 2003.

(1) A licensee or permittee shall not temporarily or permanently distribute explosive materials to any person, other than another licensee or permittee, unless he records the transaction on an explosives transaction record, ATF F 5400.4.

(2) Before the distribution of explosive materials to a nonlicensee or nonpermittee who is a resident of the State in which the licensee or permittee maintains his business premises, or to a nonlicensee or nonpermittee who is not a resident of the State in which the licensee or permittee maintains his business premises and is acquiring explosive materials under §555.105(a)(3), the licensee or permittee distributing the explosive materials shall obtain an executed ATF F 5400.4 from the distributee which contains all of the information required on the form and by the regulations in this part.

(3) Completed ATF F 5400.4 is to be retained by the licensee or permittee as part of his permanent records in accordance with paragraph (a)(4) of this section.

(4) Each ATF F 5400.4 is retained in numerical (by transaction serial number) order commencing with "1" and continuing in regular sequence. When the numbering of any series reaches "1,000,000," the licensee or permittee may recommence the series. The recommenced series is to be given an alphabetical prefix or suffix. Where there is a change in proprietorship, or in the individual, firm, corporate name or trade name, the series in use at the time of the change may be continued.

(5) The requirements of this section are in addition to any other recordkeeping requirement contained in this part.

(6) A licensee or permittee may obtain, upon request, a supply of ATF F 5400.4 from the Director.

(b) Limited Permittee Transaction Report for distribution of explosive materials on and after May 24, 2003.

(1) A licensee or permittee may not distribute explosive materials to any person who is not a licensee or permittee. A licensee or permittee may not distribute explosive materials to a limited permittee unless the distributor records the transaction on ATF Form 5400.4, Limited Permittee Transaction Report.

(2) Before distributing explosive materials to a limited permittee, the licensee or permittee must obtain an executed Form 5400.4 from the limited permittee with an original unaltered and unexpired Intrastate Purchase of Explosives Coupon (IPEC) affixed. Except when delivery of explosive materials is made by a common or contract carrier who is an agent of the limited permittee, the licensee, permittee, or an agent of the licensee or permittee, must verify the identity of the of the holder of the limited permit by examining an identification document (as defined in §555.11) and noting on the Form 5400.4 the type of document presented. The licensee or permittee must complete the appropriate section on Form 5400.4 to indicate the type and quantity of explosive materials distributed, the license or permit number of the seller, and the date of the transaction. The licensee or permittee must sign and date the form and include any other information required by the instructions on the form and the regulations in this part.

(3) One copy of Form 5400.4 must be retained by the distributor as part of his permanent records in accordance with paragraph (b)(4) of this section and for the period specified in §555.121. The distributor must mail the other copy of Form 5400.4 to the Bureau of Alcohol, Tobacco, Firearms and Explosives in accordance with the instructions on the form.

(4) Each Form 5400.4 must be retained in chronological order by date of disposition, or in alphabetical order by name of limited permittee. A licensee may not, however, use both methods in a single recordkeeping system. Where there is a change in proprietorship by a limited permittee, the forms may continue to be filed together after such change.

(5) The requirements of this section are in addition to any other recordkeeping requirement contained in this part.

(Approved by the Office of Management and Budget under control number 1140–0078)

[T.D. ATF–87, 46 FR 40384, Aug. 7, 1981, as amended by T.D. ATF–93, 46 FR 50787, Oct. 15, 1981; T.D. ATF–172, 49 FR 14941, Apr. 16, 1984; T.D. ATF–446, 66 FR 16602, Mar. 27, 2001; ATF No. 1, 68 FR 13791, Mar. 20, 2003]

§ 555.127 Daily summary of magazine transactions.

In taking the inventory required by §§555.122, 555.123, 555.124, and 555.125, a licensee or permittee shall enter the inventory in a record of daily summary transactions to be kept at each magazine of an approved storage facility; however, these records may be kept at one central location on the business premises if separate records of daily transactions are kept for each magazine. Not later than the close of the next business day, each licensee and permittee shall record by manufacturer's name or brand name, the total quantity received in and removed from each magazine during the day, and the total remaining on hand at the end of the day. Quantity entries for display fireworks may be expressed as the number and size of individual display fireworks in a finished state or as the number of packaged display segments or packaged displays. Information as to the number and size of display fireworks contained in any one packaged display segment or packaged display shall be provided to any ATF officer on request. Any discrepancy which might indicate a theft or loss of explosive materials is to be reported in accordance with §555.30.

[T.D. ATF–293, 55 FR 3722, Feb. 5, 1990, as amended by T.D. ATF–400, 63 FR 45003, Aug. 24, 1998]

§ 555.128 Discontinuance of business.

Where an explosive materials business or operations is discontinued and succeeded by a new licensee or new permittee, the records prescribed by this subpart shall appropriately reflect such facts and shall be delivered to the successor. Where discontinuance of the business or operations is absolute, the records required by this subpart must be delivered within 30 days following the business or operations discontinuance to any ATF office located in the region in which the business was located, or to the ATF Out-of-Business Records Center, Spring Mills Office Park, 882 T.J. Jackson Drive, Falling Waters, West Virginia 25419. Where State law or local ordinance requires the delivery of records to other responsible authority, the Chief, Firearms and Explosives Licensing Center may arrange for the delivery of the records required by this subpart to such authority. (See also, §555.61.)

[T.D. ATF–290, 54 FR 53054, Dec. 27, 1989, as amended by T.D. ATF–446a, 66 FR 19089, Apr. 13, 2001; ATF No. 1, 68 FR 13792, Mar. 20, 2003]

§ 555.129 Exportation.

Exportation of explosive materials is to be in accordance with the applicable provisions of section 38 of the Arms Export Control Act (22 U.S.C. 2778) and implementing regulations. However, a licensed importer, licensed manufacturer, or licensed dealer exporting explosive materials shall maintain records showing the manufacture or acquisition of explosive materials as required by this part and records showing the quantity, the manufacturer's name or brand name of explosive materials, the name and address of the foreign consignee of the explosive materials, and the date the explosive materials were exported. See §555.180 for regulations concerning the exportation of plastic explosives.

[T.D. ATF–87, 46 FR 40384, Aug. 7, 1981, as amended by T.D. ATF–387, 62 FR 8377, Feb. 25, 1997]

§ 555.130 [Reserved]

Subpart H—Exemptions

§ 555.141 Exemptions.

(a) General. Except for the provisions of §§555.180 and 555.181, this part does not apply to:

(1) Any aspect of the transportation of explosive materials via railroad, water, highway, or air which is regulated by the U.S. Department of Transportation and its agencies, and which pertains to safety. For example, regulations issued by the Department of Transportation addressing the security risk of aliens transporting explosives by commercial motor or railroad carrier from Canada preclude the enforcement of 18 U.S.C. 842(i)(5) against persons shipping, transporting, receiving, or possessing explosives incident to and in connection with the commercial transportation of explosives by truck or rail from Canada into the United States. Questions concerning this exception should be directed to ATF's Public Safety Branch in Washington, DC.

(2) The use of explosive materials in medicines and medicinal agents in the forms prescribed by the official United States Pharmacopeia or the National Formulary. "The United States Pharmacopeia and The National Formulary," USP and NF Compendia, are available from the United States Pharmacopeial Convention, Inc., 12601 Twinbrook Parkway, Rockville, Maryland 20852.

(3) The transportation, shipment, receipt, or importation of explosive materials for delivery to any agency of the United States or to any State or its political subdivision.

(4) Small arms ammunition and components of small arms ammunition.

(5) The manufacture under the regulation of the military department of the United States of explosive materials for, or their distribution to or storage or possession by, the military or naval services or other agencies of the United States.

(6) Arsenals, navy yards, depots, or other establishments owned by, or operated by or on behalf of, the United States.

(7) The importation, distribution, and storage of fireworks classified as UN0336, UN0337, UN0431, or UN0432 explosives by the U.S. Department of Transportation at 49 CFR 172.101 and generally known as "consumer fireworks" or "articles pyrotechnic."

(8) Gasoline, fertilizers, propellant actuated devices, or propellant actuated industrial tools manufactured, imported, or distributed for their intended purposes.

(9) Industrial and laboratory chemicals which are intended for use as reagents and which are packaged and shipped pursuant to U.S. Department of Transportation regulations, 49 CFR Parts 100 to 177, which do not require explosives hazard warning labels.

(10) Model rocket motors that meet all of the following criteria—

(i) Consist of ammonium perchlorate composite propellant, black powder, or other similar low explosives;

(ii) Contain no more than 62.5 grams of total propellant weight; and

(iii) Are designed as single-use motors or as reload kits capable of reloading no more than 62.5 grams of propellant into a reusable motor casing.

(b) Black powder. Except for the provisions applicable to persons required to be licensed under subpart D, this part does not apply with respect to commercially manufactured black powder in quantities not to exceed 50 pounds, percussion caps, safety and pyrotechnic fuses, quills, quick and slow matches, and friction primers, if the black powder is intended to be used solely for sporting, recreational, or cultural purposes in antique firearms, as defined in 18 U.S.C. 921(a)(16) or antique devices, as exempted from the term "destructive devices" in 18 U.S.C. 921(a)(4).

[T.D. ATF–87, 46 FR 40384, Aug. 7, 1981 as amended by T.D. ATF–87, 46 FR 46916, Sept. 23, 1981; T.D. ATF–293, 55 FR 3722, Feb. 5, 1990; T.D. ATF–387, 62 FR 8377, Feb. 25, 1997; T.D. ATF–400, 63 FR 45003, Aug. 24, 1998; ATF No. 1, 68 FR 13792, Mar. 20, 2003; ATF 6F, 71 FR 46101, Aug. 11, 2006]

§ 555.142 Relief from disabilities (effective January 24, 2003).

(a) Any person prohibited from shipping or transporting any explosive in or affecting interstate or foreign commerce or from receiving or possessing any explosive which has been shipped or transported in or affecting interstate or foreign commerce may make application for relief from disabilities under section 845(b) of the Act.

(b) An application for relief from disabilities must be filed with the Director by submitting ATF Form 5400.29, Application for Restoration of Explosives Privileges, in accordance with the instructions on the form. The application must be supported by appropriate data, including the information specified in paragraph (f) of this section. Upon receipt of an incomplete or improperly executed application for relief, the applicant will be notified of the deficiency in the application. If the application is not corrected and returned within 30 days following the date of notification, the application will be considered abandoned.

(c) (1) The Director may grant relief to an applicant if it is established to the satisfaction of the Director that the circumstances regarding the disability and the applicant's record and reputation are such that the applicant will not be likely to act in a manner dangerous to public safety and that the granting of such relief is not contrary to the public interest.

(2) Except as provided in paragraph (c)(3) of this section, the Director will not grant relief if the applicant—

(i) Has not been discharged from parole or probation for a period of at least 2 years;

(ii) Is a fugitive from justice;

(iii) Is a prohibited alien;

(iv) Is an unlawful user of or addicted to any controlled substance;

(v) Has been adjudicated a mental defective or committed to a mental institution, unless the applicant was subsequently determined by a court, board, commission, or other lawful authority to have been restored to mental competency, to be no longer suffering from a mental disorder, and to have had all rights restored; or

(vi) Is prohibited by the law of the State where the applicant resides from receiving or possessing explosive materials.

(3)(i) The Director may grant relief to aliens who have been lawfully admitted to the United States or to persons who have not been discharged from parole or probation for a period of at least 2 years if he determines that the applicant has a compelling need to possess explosives, such as for purposes of employment.

(ii) The Director may grant relief to the persons identified in paragraph (c)(2) of this section in extraordinary circumstances where the granting of such relief is consistent with the public interest.

(d) A person who has been granted relief under this section is relieved of all disabilities imposed by the Act for the disabilities disclosed in the application. The granting of relief will not affect any disabilities incurred subsequent to the date the application was filed. Relief from disabilities granted to aliens will be effective only so long as the alien retains his or her lawful immigration status.

(e)(1) A licensee or permittee who is under indictment or information for, or convicted of, a crime punishable by imprisonment for a term exceeding one year during the term of a current license or permit, or while he has pending a license or permit renewal application, shall not be barred from licensed or permit operations for 30 days after the date of indictment or information or 30 days after the date upon which his conviction becomes final. Also, if he files his application for relief under this section within such 30 day period, he may further continue licensed or permit operations while his application is pending. A licensee or permittee who does not file an application within 30 days from the date of his indictment or information, or within 30 days from the date his conviction becomes final, shall not continue licensed or permit operations beyond 30 days from the date of his indictment or information or beyond 30 days from the date his conviction becomes final.

(2) In the event the term of a license or permit of a person expires during the 30 day period following the date of indictment of information of during the 30 day period after the date upon which his conviction becomes final or while his application for relief is pending, he shall file a timely application for renewal of his license or permit in order to continue licensed or permit operations. The license or permit application is to show that the applicant has been indicted or under information for, or convicted of, a crime punishable by imprisonment for a term exceeding one year.

(3) A licensee or permittee shall not continue licensed or permit operations beyond 30 days following the date the Director issues notification that the licensee's or permittee's application for removal of the disabilities resulting from an indictment, information or conviction has been denied.

(4) When a licensee or permittee may no longer continue licensed or permit operations under this section, any application for renewal of license of permit filed by the licensee or permittee while his application for removal of disabilities resulting from an indictment, information or conviction is pending, will be denied by the regional director (compliance).

(f)(1) Applications for relief from disabilities must include the following information:

(i) In the case of a corporation, or of any person having the power to direct or control the management of the corporation, information as to the absence of culpability in the offense for which the corporation, or any such person, was indicted, formally accused or convicted;

(ii) In the case of an applicant who is an individual, two properly completed FBI Forms FD–258 (fingerprint card), and a written statement from each of three references who are not related to the applicant by blood or marriage and have known the applicant for at least 3 years, recommending the granting of relief;

(iii) Written consent to examine and obtain copies of records and to receive statements and information regarding the applicant's background, including records, statements and other information concerning employment, medical history, military service, immigration status, and criminal record;

(iv) In the case of an applicant having been convicted of a crime punishable by imprisonment for a term exceeding one year, a copy of the indictment or information on which the applicant was convicted, the judgment of conviction or record of any plea of nolo contendere or plea of guilty or finding of guilt by the court;

(v) In the case of an applicant under indictment, a copy of the indictment or information;

(vi) In the case of an applicant who has been adjudicated a mental defective or committed to a mental institution, a copy of the order of a court, board, commission, or other lawful authority that made the adjudication or ordered the commitment, any petition that sought to have the applicant so adjudicated or committed, any medical records reflecting the reasons for commitment and diagnoses of the applicant, and any court order or finding of a court, board, commission, or other lawful authority showing the applicant's discharge from commitment, restoration of mental competency and the restoration of rights;

(vii) In the case of an applicant who has been discharged from the Armed Forces under dishonorable conditions, a copy of the applicant's Certificate of Release or Discharge from Active Duty (Department of Defense Form 214), Charge Sheet (Department of Defense Form 458), and final court martial order;

(viii) In the case of an applicant who, having been a citizen of the United States, has renounced his or her citizenship, a copy of the formal renunciation of nationality before a diplomatic or consular officer of the United States in a foreign state or before an officer designated by the Attorney General when the United States was in a state of war (see 8 U.S.C. 1481(a)(5) and (6)); and

(ix) In the case of an applicant who is an alien, documentation that the applicant is an alien who has been lawfully admitted to the United States; certification from the applicant including the applicant's INS-issued alien number or admission number, country/countries of citizenship, and immigration status, and certifying that the applicant is legally authorized to work in the United States, or other purposes for which possession of explosives is required; certification from an appropriate law enforcement agency of the applicant's country of citizenship stating that the applicant does not have a criminal record; and, if applicable, certification from a Federal explosives licensee or permittee or other employer stating that the applicant is employed by the employer and must possess explosive materials for purposes of employment. These certifications must be submitted in English.

(2) Any record or document of a court or other government entity or official required by paragraph (f)(1) of this section must be certified by the court or other government entity or official as a true copy.

(Approved by the Office of Management and Budget under control number 1140–0076)

[T.D. ATF–87, 46 FR 40384, Aug. 7, 1981. Redesignated by T.D. ATF–487, 68 FR 3748, Jan. 24, 2003. ATF No. 1, 68 FR 13792, Mar. 20, 2003]

Subpart I—Unlawful Acts, Penalties, Seizures and Forfeitures

§ 555.161 Engaging in business without a license.

Any person engaging in the business of importing, manufacturing, or dealing in explosive materials without a license issued under the Act, shall be fined not more than $10,000 or imprisoned not more than 10 years, or both.

§ 555.162 False statement or representation.

Any person who knowingly withholds information or makes any false or fictitious oral or written statement or furnishes or exhibits any false, fictitious, or misrepresented identification, intended or likely to deceive for the purpose of obtaining explosive materials, or a license, permit, exemption, or relief from disability under the Act, shall be fined not more than $10,000 or imprisoned not more than 10 years, or both.

§ 555.163 False entry in record.

Any licensed importer, licensed manufacturer, licensed dealer, or permittee who knowingly makes any false entry in any record required to be kept under subpart G of this part, shall be fined not more than $10,000 or imprisoned not more than 10 years, or both.

[T.D. ATF–87, 46 FR 40384, Aug. 7, 1981, as amended by T.D. ATF–400, 63 FR 45003, Aug. 24, 1998]

§ 555.164 Unlawful storage.

Any person who stores any explosive material in a manner not in conformity with this part, shall be fined not more than $1,000 or imprisoned not more than one year, or both.

§ 555.165 Failure to report theft or loss.

(a) Any person who has knowledge of the theft or loss of any explosive materials from his stock and fails to report the theft or loss within 24 hours of discovery in accordance with §555.30, shall be fined not more than $1,000 or imprisoned not more than one year, or both.

(b) On and after January 24, 2003, any licensee or permittee who fails to report a theft of explosive materials in accordance with §555.30 will be fined under title 18 U.S.C., imprisoned not more than 5 years, or both.

[T.D. ATF–87, 46 FR 40384, Aug. 7, 1981, as amended by ATF No. 1, 68 FR 13793, Mar. 20, 2003]

§ 555.166 Seizure or forfeiture.

Any explosive materials involved or used or intended to be used in any violation of the Act or of this part or in any violation of any criminal law of the United States are subject to seizure and forfeiture, and all provisions of title 26, U.S.C. relating to the seizure, forfeiture, and disposition of firearms, as defined in 26 U.S.C. 5845(a), will, so far as applicable, extend to seizures and forfeitures under the Act. (See §72.27 of this title for regulations on summary destruction of explosive materials which are impracticable or unsafe to remove to a place of storage.)

[T.D. ATF–87, 46 FR 40384, Aug. 7, 1981, as amended by T.D. ATF–363, 60 FR 17449, Apr. 6, 1995]

Subpart J—Marking of Plastic Explosives

§ 555.180 Prohibitions relating to unmarked plastic explosives.

(a) No person shall manufacture any plastic explosive that does not contain a detection agent.

(b) No person shall import or bring into the United States, or export from the United States, any plastic explosive that does not contain a detection agent. This paragraph does not apply to the importation or bringing into the United States, or the exportation from the United States, of any plastic explosive that was imported or brought into, or manufactured in the United States prior to April 24, 1996, by or on behalf of any agency of the United States performing military or police functions (including any military reserve component) or by or on behalf of the National Guard of any State, not later than 15 years after the date of entry into force of the Convention on the Marking of Plastic Explosives with respect to the United States, i.e., not later than June 21, 2013.

(c) No person shall ship, transport, transfer, receive, or possess any plastic explosive that does not contain a detection agent. This paragraph does not apply to:

(1) The shipment, transportation, transfer, receipt, or possession of any plastic explosive that was imported or brought into, or manufactured in the United States prior to April 24, 1996, by any person during the period beginning on that date and ending on April 24, 1999; or

(2) The shipment, transportation, transfer, receipt, or possession of any plastic explosive that was imported or brought into, or manufactured in the United States prior to April 24, 1996, by or on behalf of any agency of the United States performing a military or police function (including any military reserve component) or by or on behalf of the National Guard of any State, not later than 15 years after the date of entry into force of the Convention on the Marking of Plastic Explosives with respect to the United States, i.e., not later than June 21, 2013.

(d) When used in this subpart, terms are defined as follows:

(1) Convention on the Marking of Plastic Explosives means the Convention on the Marking of Plastic Explosives for the Purposes of Detection, Done at Montreal on 1 March 1991.

(2) "Date of entry into force" of the Convention on the Marking of Plastic Explosives means that date on which the Convention enters into force with respect to the U.S. in accordance with the provisions of Article XIII of the Convention on the Marking of Plastic Explosives. The Convention entered into force on June 21, 1998.

(3) Detection agent means any one of the substances specified in this paragraph when introduced into a plastic explosive or formulated in such explosive as a part of the manufacturing process in such a manner as to achieve homogeneous distribution in the finished explosive, including—

(i) Ethylene glycol dinitrate (EGDN), C[2]H[4](NO[3])[2], molecular weight 152, when the minimum concentration in the finished explosive is 0.2 percent by mass;

(ii) 2,3-Dimethyl-2,3-dinitrobutane (DMNB), C[6]H[12](NO[2])[2], molecular weight 176, when the minimum concentration in the finished explosive is 0.1 percent by mass;

(iii) Para-Mononitrotoluene (p-MNT), C[7]H[7]NO[2], molecular weight 137, when the minimum concentration in the finished explosive is 0.5 percent by mass;

(iv) Ortho-Mononitrotoluene (o-MNT), C[7]H[7]NO[2], molecular weight 137, when the minimum concentration in the finished explosive is 0.5 percent by mass; and

(v) Any other substance in the concentration specified by the Director, after consultation with the Secretary of State and Secretary of Defense, that has been added to the table in Part 2 of the Technical Annex to the Convention on the Marking of Plastic Explosives.

(4) Plastic explosive means an explosive material in flexible or elastic sheet form formulated with one or more high explosives which in their pure form has a vapor pressure less than 10<4>Pa at a temperature of 25 degrees C, is formulated with a binder material, and is as a mixture malleable or flexible at normal room temperature. High explosives, as defined in §555.202(a), are explosive materials which can be caused to detonate by means of a blasting cap when unconfined.

[T.D. ATF–387, 62 FR 8376, Feb. 25, 1997, as amended by T.D. ATF–419, 64 FR 55628, Oct. 14, 1999]

§ 555.181 Reporting of plastic explosives.

All persons, other than an agency of the United States (including any military reserve component) or the National Guard of any State, possessing any plastic explosive on April 24, 1996, shall submit a report to the Director no later than August 22, 1996. The report shall be in writing and mailed by certified mail (return receipt requested) to the Director at P.O. Box 50204, Washington, DC 20091–0204. The report shall include the quantity of plastic explosives possessed on April 24, 1996; any marks of identification on such explosives; the name and address of the manufacturer or importer; the storage location of such explosives, including the city and State; and the name and address of the person possessing the plastic explosives.

(Approved by the Office of Management and Budget under control number 1512–0535)

[T.D. ATF–382, 61 FR 38085, July 23, 1996, as amended by T.D. ATF–387, 62 FR 8377, Feb. 25, 1997]

§ 555.182 Exceptions.

It is an affirmative defense against any proceeding involving §§555.180 and 555.181 if the proponent proves by a preponderance of the evidence that the plastic explosive—

(a) Consisted of a small amount of plastic explosive intended for and utilized solely in lawful—

(1) Research, development, or testing of new or modified explosive materials;

(2) Training in explosives detection or development or testing of explosives detection equipment; or

(3) Forensic science purposes; or

(b) Was plastic explosive that, by April 24, 1999, will be or is incorporated in a military device within the territory of the United States and remains an integral part of such military device, or is intended to be, or is incorporated in, and remains an integral part of a military device that is intended to become, or has become, the property of any agency of the United States performing military or police functions (including any military reserve component) or the National Guard of any State, wherever such device is located. For purposes of this paragraph, the term "military device" includes, but is not restricted to, shells, bombs, projectiles, mines, missiles, rockets, shaped charges, grenades, perforators, and similar devices lawfully manufactured exclusively for military or police purposes.
[T.D. ATF–387, 62 FR 8377, Feb. 25, 1997]

§ 555.183 Importation of plastic explosives on or after April 24, 1997.

Persons filing Form 6 applications for the importation of plastic explosives on or after April 24, 1997, shall attach to the application the following written statement, prepared in triplicate, executed under the penalties of perjury:

(a) "I declare under the penalties of perjury that the plastic explosive to be imported contains a detection agent as required by 27 CFR 555.180(b)"; or

(b) "I declare under the penalties of perjury that the plastic explosive to be imported is a "small amount" to be used for research, training, or testing purposes and is exempt from the detection agent requirement pursuant to 27 CFR 555.182."
[T.D. ATF–387, 62 FR 8377, Feb. 25, 1997]

§ 555.184 Statements of process and samples.

(a) A complete and accurate statement of process with regard to any plastic explosive or to any detection agent that is to be introduced into a plastic explosive or formulated in such plastic explosive shall be submitted by a licensed manufacturer or licensed importer, upon request, to the Director.

(b) Samples of any plastic explosive or detection agent shall be submitted by a licensed manufacturer or licensed importer, upon request, to the Director.

(Paragraph (a) approved by the Office of Management and Budget under control number 1512–0539)
[T.D. ATF–387, 62 FR 8378, Feb. 25, 1997]

§ 555.185 Criminal sanctions.

Any person who violates the provisions of 18 U.S.C. 842(l)–(o) shall be fined under title 18, U.S.C., imprisoned for not more than 10 years, or both.
[T.D. ATF–387, 62 FR 8378, Feb. 25, 1997]

§ 555.186 Seizure or forfeiture.

Any plastic explosive that does not contain a detection agent in violation of 18 U.S.C. 842(l)–(n) is subject to seizure and forfeiture, and all provisions of 19 U.S.C. 1595a, relating to seizure, forfeiture, and disposition of merchandise introduced or attempted to be introduced into the U.S. contrary to law, shall extend to seizures and forfeitures under this subpart. See §72.27 of this chapter for regulations on summary destruction of plastic explosives that do not contain a detection agent.
[T.D. ATF–387, 62 FR 8378, Feb. 25, 1997]

Subpart K—Storage

§ 555.201 General.

(a) Section 842(j) of the Act and §555.29 of this part require that the storage of explosive materials by any person must be in accordance with the regulations in this part. Further, section 846 of this Act authorizes regulations to prevent the recurrence of accidental explosions in which explosive materials were involved. The storage standards prescribed by this subpart confer no right or privileges to store explosive materials in a manner contrary to State or local law.

(b) The Director may authorize alternate construction for explosives storage magazines when it is shown that the alternate magazine construction is substantially equivalent to the standards of safety and security contained in this subpart. Any alternate explosive magazine construction approved by the Director prior to August 9, 1982, will continue as approved unless notified in writing by the Director. Any person intending to use alternate magazine construction shall submit a letter application to the regional director (compliance) for transmittal to the Director, specifically describing the proposed magazine. Explosive materials may not be stored in alternate magazines before the applicant has been notified that the application has been approved.

(c) A licensee or permittee who intends to make changes in his magazines, or who intends to construct or acquire additional magazines, shall comply with §555.63.

(d) The regulations set forth in §§555.221 through 555.224 pertain to the storage of display fireworks, pyrotechnic compositions, and explosive materials used in assembling fireworks and articles pyrotechnic.

(e) The provisions of §555.202(a) classifying flash powder and bulk salutes as high explosives are mandatory after March 7, 1990: Provided, that those persons who hold licenses or permits under this part on that date shall, with respect to the premises covered by such licenses or permits, comply with the high explosives storage requirements for flash powder and bulk salutes by March 7, 1991.

(f) Any person who stores explosive materials shall notify the authority having jurisdiction for fire safety in the locality in which the explosive materials are being stored of the type, magazine capacity, and location of each site where such explosive materials are stored. Such notification shall be made orally before the end of the day on which storage of the explosive materials commenced and in writing within 48 hours from the time such storage commenced.

(Paragraph (f) approved by the Office of Management and Budget under control number 1512–0536)

[T.D. ATF–87, 46 FR 40384, Aug. 7, 1981, as amended by T.D. ATF–293, 55 FR 3722, Feb. 5, 1990; T.D. ATF–400, 63 FR 45003, Aug. 24, 1998]

§ 555.202 Classes of explosive materials.

For purposes of this part, there are three classes of explosive materials. These classes, together with the description of explosive materials comprising each class, are as follows:

(a) High explosives. Explosive materials which can be caused to detonate by means of a blasting cap when unconfined, (for example, dynamite, flash powders, and bulk salutes). See also §555.201(e).

(b) Low explosives. Explosive materials which can be caused to deflagrate when confined (for example, black powder, safety fuses, igniters, igniter cords, fuse lighters, and "display fireworks" classified as UN0333, UN0334, or UN0335 by the U.S. Department of Transportation regulations at 49 CFR 172.101, except for bulk salutes).

(c) Blasting agents. (For example, ammonium nitrate-fuel oil and certain water-gels (see also §555.11).

[T.D. ATF–87, 46 FR 40384, Aug. 7, 1981, as amended by T.D. ATF–293, 55 FR 3722, Feb. 5, 1990; T.D. ATF–400, 63 FR 45003, Aug. 24, 1998]

§ 555.203 Types of magazines.

For purposes of this part, there are five types of magazines. These types, together with the classes of explosive materials, as defined in §555.202, which will be stored in them, are as follows:

(a) Type 1 magazines. Permanent magazines for the storage of high explosives, subject to the limitations prescribed by §§555.206 and 555.213. Other classes of explosive materials may also be stored in type 1 magazines.

(b) Type 2 magazines. Mobile and portable indoor and outdoor magazines for the storage of high explosives, subject to the limitations prescribed by §§555.206, 555.208(b), and 555.213. Other classes of explosive materials may also be stored in type 2 magazines.

(c) Type 3 magazines. Portable outdoor magazines for the temporary storage of high explosives while attended (for example, a "day-box"), subject to the limitations prescribed by §§555.206 and 555.213. Other classes of explosives materials may also be stored in type 3 magazines.

(d) Type 4 magazines. Magazines for the storage of low explosives, subject to the limitations prescribed by §§555.206(b), 555.210(b), and 555.213. Blasting agents may be stored in type 4 magazines, subject to the limitations prescribed by §§555.206(c), 555.211(b), and 555.213. Detonators that will not mass detonate may also be stored in type 4 magazines, subject to the limitations prescribed by §§555.206(a), 555.210(b), and 555.213.

(e) Type 5 magazines. Magazines for the storage of blasting agents, subject to the limitations prescribed by §§555.206(c), 555.211(b), and 555.213.

§ 555.204 Inspection of magazines.

Any person storing explosive materials shall inspect his magazines at least every seven days. This inspection need not be an inventory, but must be sufficient to determine whether there has been unauthorized entry or attempted entry into the magazines, or unauthorized removal of the contents of the magazines.

§ 555.205 Movement of explosive materials.

All explosive materials must be kept in locked magazines meeting the standards in this subpart unless they are:

(a) In the process of manufacture;

(b) Being physically handled in the operating process of a licensee or user;

(c) Being used; or

(d) Being transported to a place of storage or use by a licensee or permittee or by a person who has lawfully acquired explosive materials under §555.106.

§ 555.206 Location of magazines.

(a) Outdoor magazines in which high explosives are stored must be located no closer to inhabited buildings, passenger railways, public highways, or other magazines in which high explosives are stored, than the minimum distances specified in the table of distances for storage of explosive materials in §555.218.

(b) Outdoor magazines in which low explosives are stored must be located no closer to inhibited buildings, passenger railways, public highways, or other magazines in which explosive materials are stored, than the minimum distances specified in the table of distances for storage of low explosives in §555.219, except that the table of distances in §555.224 shall apply to the storage of display fireworks. The distances shown in §555.219 may not be reduced by the presence of barricades.

(c)(1) Outdoor magazines in which blasting agents in quantities of more than 50 pounds are stored must be located no closer to inhabited buildings, passenger railways, or public highways than the minimum distances specified in the table of distances for storage of explosive materials in §555.218.

(2) Ammonium nitrate and magazines in which blasting agents are stored must be located no closer to magazines in which high explosives or other blasting agents are stored than the minimum distances specified in the table of distances for the separation of ammonium nitrate and blasting agents in §555.220. However, the minimum distances for magazines in which explosives and blasting agents are stored from inhabited buildings, etc., may not be less than the distances specified in the table of distances for storage of explosives materials in §555.218.

[T.D. ATF–87, 46 FR 40384, Aug. 7, 1981, as amended by T.D. ATF–293, 55 FR 3722, Feb. 5, 1990; T.D. ATF–400, 63 FR 45003, Aug. 24, 1998]

§ 555.207 Construction of type 1 magazines.

A type 1 magazine is a permanent structure: a building, an igloo or "Army-type structure", a tunnel, or a dugout. It is to be bullet-resistant, fire-resistant, weather-resistant, theft-resistant, and ventilated.

(a) Buildings. All building type magazines are to be constructed of masonry, wood, metal, or a combination of these materials, and have no openings except for entrances and ventilation. The ground around building magazines must slope away for drainage or other adequate drainage provided.

(1) Masonry wall construction. Masonry wall construction is to consist of brick, concrete, tile, cement block, or cinder block and be not less than 6 inches in thickness. Hollow masonry units used in construction must have all hollow spaces filled with well-tamped, coarse, dry sand or weak concrete (at least a mixture of one part cement and eight parts of sand with enough water to dampen the mixture while tamping in place). Interior walls are to be constructed of, or covered with, a nonsparking material.

(2) Fabricated metal wall construction. Metal wall construction is to consist of sectional sheets of steel or aluminum not less than number 14-gauge, securely fastened to a metal framework. Metal wall construction is either lined inside with brick, solid cement blocks, hardwood not less than four inches thick, or will have at least a six inch sand fill between interior and exterior walls. Interior walls are to be constructed of, or covered with, a nonsparking material.

(3) Wood frame wall construction. The exterior of outer wood walls is to be covered with iron or aluminum not less than number 26-gauge. An inner wall of, or covered with nonsparking material will be constructed so as to provide a space of not less than six inches between the outer and inner walls. The space is to be filled with coarse, dry sand or weak concrete.

(4) Floors. Floors are to be constructed of, or covered with, a nonsparking material and shall be strong enough to bear the weight of the maximum quantity to be stored. Use of pallets covered with a nonsparking material is considered equivalent to a floor constructed of or covered with a nonsparking material.

(5) Foundations. Foundations are to be constructed of brick, concrete, cement block, stone, or wood posts. If piers or posts are used, in lieu of a continuous foundation, the space under the buildings is to be enclosed with metal.

(6) Roof. Except for buildings with fabricated metal roofs, the outer roof is to be covered with no less than number 26-guage iron or aluminum, fastened to at least ⅞ inch sheathing.

(7) Bullet-resistant ceilings or roofs. Where it is possible for a bullet to be fired directly through the roof and into the magazine at such an angle that the bullet would strike the explosives within, the magazine is to be protected by one of the following methods:

(i) A sand tray lined with a layer of building paper, plastic, or other nonporous material, and filled with not less than four inches of coarse, dry sand, and located at the tops of inner walls covering the entire ceiling area, except that portion necessary for ventilation.

(ii) A fabricated metal roof constructed of $^{3}/_{16}$-inch plate steel lined with four inches of hardwood. (For each additional ⅙ inch of plate steel, the hardwood lining may be decreased one inch.)

(8) Doors. All doors are to be constructed of not less than ¼ inch plate steel and lined with at least two inches of hardwood. Hinges and hasps are to be attached to the doors by welding, riveting or bolting (nuts on inside of door). They are to be installed in such a manner that the hinges and hasps cannot be removed when the doors are closed and locked.

(9) Locks. Each door is to be equipped with (i) two mortise locks; (ii) two padlock fastened in separate hasps and staples; (iii) a combination of a mortise lock and a padlock; (iv) a mortise lock that requires two keys to open; or (v) a three-point lock. Padlocks must have at least five tumblers and a casehardened shackle of at least ⅜ inch diameter. Padlocks must be protected with not less than ¼ inch steel hoods constructed so as to prevent sawing or lever action on the locks, hasps, and staples. These requirements do not apply to magazine doors that are adequately secured on the inside by means of a bolt, lock, or bar that cannot be actuated from the outside.

(10) Ventilation. Ventilation is to be provided to prevent dampness and heating of stored explosive materials. Ventilation openings must be screened to prevent the entrance of sparks. Ventilation openings in side walls and foundations must be offset or shielded for bullet-resistant purposes. Magazines having foundation and roof ventilators with the air circulating between the side walls and the floors and between the side walls and the ceiling must have a wooden lattice lining or equivalent to prevent the packages of explosive materials from being stacked against the side walls and blocking the air circulation.

(11) Exposed metal. No sparking material is to be exposed to contact with the stored explosive materials. All ferrous metal nails in the floor and side walls, which might be exposed to contact with explosive materials, must be blind nailed, countersunk, or covered with a nonsparking lattice work or other nonsparking material.

(b) Igloos, "Army-type structures", tunnels, and dugouts. Igloo, "Army-type structure", tunnel, and dugout magazines are to be constructed of reinforced concrete, masonry, metal, or a combination of these materials. They must have an earthmound covering of not less than 24 inches on the top, sides and rear unless the magazine meets the requirements of paragraph (a)(7) of this section. Interior walls and floors must be constructed of, or covered with, a nonsparking material. Magazines of this type are also to be constructed in conformity with the requirements of paragraph (a)(4) and paragraphs (a)(8) through (11) of this section.

§ 555.208 Construction of type 2 magazines.

A type 2 magazine is a box, trailer, semitrailer, or other mobile facility.

(a) Outdoor magazines—

(1) General. Outdoor magazines are to be bullet-resistant, fire-resistant, weather-resistant, theft-resistant, and ventilated. They are to be supported to prevent direct contact with the ground and, if less than one cubic yard in size, must be securely fastened to a fixed object. The ground around outdoor magazines must slope away for drainage or other adequate drainage provided. When unattended, vehicular magazines must have wheels removed or otherwise effectively immobilized by kingpin locking devices or other methods approved by the Director.

(2) Exterior construction. The exterior and doors are to be constructed of not less than ¼-inch steel and lined with at least two inches of hardwood. Magazines with top openings will have lids with water-resistant seals or which overlap the sides by at least one inch when in a closed position.

(3) Hinges and hasps. Hinges and hasps are to be attached to doors by welding, riveting, or bolting (nuts on inside of door). Hinges and hasps must be installed so that they cannot be removed when the doors are closed and locked.

(4) Locks. Each door is to be equipped with (i) two mortise locks; (ii) two padlocks fastened in separate hasps and staples; (iii) a combination of a mortise lock and a padlock; (iv) a mortise lock that requires two keys to open; or (v) a three-point lock. Padlocks must have at least five tumblers and a case-hardened shackle of at least ⅜-inch diameter. Padlocks must be protected with not less than ¼-inch steel hoods constructed so as to prevent sawing or lever action on the locks, hasps, and staples. These requirements do not apply to magazine doors that are adequately secured on the inside by means of a bolt, lock, or bar that cannot be actuated from the outside.

(b) Indoor magazines—

(1) General. Indoor magazines are to be fire-resistant and theft-resistant. They need not be bullet-resistant and weather-resistant if the buildings in which they are stored provide protection from the weather and from bullet penetration. No indoor magazine is to be located in a residence or dwelling. The indoor storage of high explosives must not exceed a quantity of 50 pounds. More than one indoor magazine may be located in the same building if the total quantity of explosive materials stored does not exceed 50 pounds. Detonators must be stored in a separate magazine (except as provided in §555.213) and the total quantity of detonators must not exceed 5,000.

(2) Exterior construction. Indoor magazines are to be constructed of wood or metal according to one of the following specifications:

(i) Wood indoor magazines are to have sides, bottoms and doors constructed of at least two inches of hardwood and are to be well braced at the corners. They are to be covered with sheet metal of not less than number 26-gauge (.0179 inches). Nails exposed to the interior of magazines must be countersunk.

(ii) Metal indoor magazines are to have sides, bottoms and doors constructed of not less than number 12-gauge (.1046 inches) metal and be lined inside with a nonsparking material. Edges of metal covers must overlap sides at least one inch.

(3) Hinges and hasps. Hinges and hasps are to be attached to doors by welding, riveting, or bolting (nuts on inside of door). Hinges and hasps must be installed so that they cannot be removed when the doors are closed and locked.

(4) Locks. Each door is to be equipped with (i) two mortise locks; (ii) two padlocks fastened in separate hasps and staples; (iii) a combination of a mortise lock and a padlock; (iv) a mortise lock that requires two keys to open; or (v) a three-point lock. Padlocks must have at least five tumblers and a case-hardened shackle of at least ⅜-inch diameter. Padlocks must be protected with not less than ¼-inch steel hoods constructed so as to prevent sawing or lever action on the locks, hasps, and staples. Indoor magazines located in secure rooms that are locked as provided in this subparagraph may have each door locked with one steel padlock (which need not be protected by a steel hood) having at least five tumblers and a case-hardened shackle of at least ⅜-inch diameter, if the door hinges and lock hasp are securely fastened to the magazine.

These requirements do not apply to magazine doors that are adequately secured on the inside by means of a bolt, lock, or bar that cannot be actuated from the outside.

(c) Detonator boxes. Magazines for detonators in quantities of 100 or less are to have sides, bottoms and doors constructed of not less than number 12-gauge (.1046 inches) metal and lined with a nonsparking material. Hinges and hasps must be attached so they cannot be removed from the outside. One steel padlock (which need not be protected by a steel hood) having at least five tumblers and a case-hardened shackle of at least ⅜-inch diameter is sufficient for locking purposes.

§ 555.209 Construction of type 3 magazines.

A type 3 magazine is a "day-box" or other portable magazine. It must be fire-resistant, weather-resistant, and theft-resistant. A type 3 magazine is to be constructed of not less than number 12-gauge (.1046 inches) steel, lined with at least either ½-inch plywood or ½-inch Masonite-type hardboard.

Doors must overlap sides by at least one inch. Hinges and hasps are to be attached by welding, riveting or bolting (nuts on inside).

One steel padlock (which need not be protected by a steel hood) having at least five tumblers and a case-hardened shackle of at least ⅜-inch diameter is sufficient for locking purposes. Explosive materials are not to be left unattended in type 3 magazines and must be removed to type 1 or 2 magazines for unattended storage.

§ 555.210 Construction of type 4 magazines.

A type 4 magazine is a building, igloo or "Army-type structure", tunnel, dugout, box, trailer, or a semitrailer or other mobile magazine.

(a) Outdoor magazines—

(1) General. Outdoor magazines are to be fire-resistant, weather-resistant, and theft-resistant. The ground around outdoor magazines must slope away for drainage or other adequate drainage be provided. When unattended, vehicular magazines must have wheels removed or otherwise be effectively immobilized by kingpin locking devices or other methods approved by the Director.

(2) Construction. Outdoor magazines are to be constructed of masonry, metal-covered wood, fabricated metal, or a combination of these materials. Foundations are to be constructed of brick, concrete, cement block, stone, or metal or wood posts. If piers or posts are used, in lieu of a continuous foundation, the space under the building is to be enclosed with fire-resistant material. The walls and floors are to be constructed of, or covered with, a nonsparking material or lattice work. The doors must be metal or solid wood covered with metal.

(3) Hinges and hasps. Hinges and hasps are to be attached to doors by welding, riveting, or bolting (nuts on inside of door). Hinges and hasps must be installed so that they cannot be removed when the doors are closed and locked.

(4) Locks. Each door is to be equipped with (i) two mortise locks; (ii) two padlocks fastened in separate hasps and staples; (iii) a combination of a mortise lock and a padlock; (iv) a mortise lock that requires two keys to open; or (v) a three-point lock. Padlocks must have at least five tumblers and case-hardened shackle of at least ⅜ inch diameter. Padlocks must be protected with not less than ¼ inch steel hoods constructed so as to prevent sawing or lever action on the locks, hasps, and staples.

These requirements do not apply to magazine doors that are adequately secured on the inside by means of a bolt, lock, or bar that cannot be actuated from the outside.

(b) Indoor magazine—

(1) General. Indoor magazines are to be fire-resistant and theft-resistant. They need not be weather-resistant if the buildings in which they are stored provide protection from the weather. No indoor magazine is to be located in a residence or dwelling. The indoor storage of low explosives must not exceed a quantity of 50 pounds. More than one indoor magazine may be located in the same building if the total quantity of explosive materials stored does not exceed 50 pounds. Detonators that will not mass detonate must be stored in a separate magazine and the total number of electric detonators must not exceed 5,000.

(2) Construction. Indoor magazines are to be constructed of masonry, metal-covered wood, fabricated metal, or a combination of these materials. The walls and floors are to be constructed of, or covered with, a nonsparking material. The doors must be metal or solid wood covered with metal.

(3) Hinges and hasps. Hinges and hasps are to be attached to doors by welding, riveting, or bolting (nuts on inside of door). Hinges and hasps must be installed so that they cannot be removed when the doors are closed and locked.

(4) Locks. Each door is to be equipped with (i) two mortise locks; (ii) two padlocks fastened in separate hasps and staples; (iii) a combination of a mortise lock and padlock; (iv) a mortise lock that requires two keys to open; or (v) a three-point lock.

Padlocks must have at least five tumblers and a case-hardened shackle of at least ⅜ inch diameter. Padlocks must be protected with not less than ¼ inch steel hoods constructed so as to prevent sawing or lever action on the locks, hasps, and staples. Indoor magazines located in secure rooms that are locked as provided in this subparagraph may have each door locked with one steel padlock (which need not be protected by a steel hood) having at least five tumblers and a case-hardened shackle of at least ⅜ inch diameter, if the door hinges and lock hasp are securely fastened to the magazine.

These requirements do not apply to magazine doors that are adequately secured on the inside by means of a bolt, lock, or bar that cannot be actuated from the outside.

§ 555.211 Construction of type 5 magazines.

A type 5 magazine is a building, igloo or "Army-type structure", tunnel, dugout, bin, box, trailer, or a semitrailer or other mobile facility.

(a) Outdoor magazines—

(1) General. Outdoor magazines are to be weather-resistant and theft-resistant. The ground around magazines must slope away for drainage or other adequate drainage be provided. When unattended, vehicular magazines must have wheels removed or otherwise be effectively immobilized by kingpin locking devices or other methods approved by the Director.

(2) Construction. The doors are to be constructed of solid wood or metal.

(3) Hinges and hasps. Hinges and hasps are to be attached to doors by welding, riveting, or bolting (nuts on inside of door). Hinges and hasps must be installed so that they cannot be removed when the doors are closed and locked.

(4) Locks. Each door is to be equipped with (i) two mortise locks; (ii) two padlocks fastened in separate hasps and staples; (iii) a combination of a mortise lock and a padlock; (iv) a mortise lock that requires two keys to open; or (v) a three-point lock. Padlocks must have at least five tumblers and a case-hardened shackle of at least ⅜ inch diameter. Padlocks must be protected with not less than ¼ inch steel hoods constructed so as to prevent sawing or lever action on the locks, hasps, and staples. Trailers, semitrailers, and similar vehicular magazines may, for each door, be locked with one steel padlock (which need not be protected by a steel hood) having at least five tumblers and a case-hardened shackle of at least ⅜ inch diameter, if the door hinges and lock hasp are securely fastened to the magazine and to the door frame.

These requirements do not apply to magazine doors that are adequately secured on the inside by means of a bolt, lock, or bar that cannot be actuated from the outside.

(5) Placards. The placards required by Department of Transportation regulations at 49 CFR part 172, subpart F, for the transportation of blasting agents shall be displayed on all magazines.

(b) Indoor magazines—

(1) General. Indoor magazines are to be theft-resistant. They need not be weather-resistant if the buildings in which they are stored provide protection from the weather. No indoor magazine is to be located in a residence or dwelling. Indoor magazines containing quantities of blasting agents in excess of 50 pounds are subject to the requirements of §555.206 of this subpart.

(2) Construction. The doors are to be constructed of wood or metal.

(3) Hinges and hasps. Hinges and hasps are to be attached to doors by welding, riveting, or bolting (nuts on inside). Hinges and hasps must be installed so that they cannot be removed when the doors are closed and locked.

(4) Locks. Each door is to be equipped with (i) two mortise locks; (ii) two padlocks fastened in separate hasps and staples; (iii) a combination of a mortise lock and a padlock; (iv) a mortise lock that requires two keys to open; or (v) a three-point lock.

Padlocks must have at least five tumblers and a case-hardened shackle of at least ⅜ inch diameter. Padlocks must be protected with not less than ¼ inch steel hoods constructed so as to prevent sawing or lever action on the locks, hasps, and staples. Indoor magazines located in secure rooms that are locked as provided in this subparagraph may have each door locked with one steel padlock (which need not be protected by a steel hood) having at least five tumblers and a case-hardened shackle of at least ⅜ inch diameter, if the door hinges and lock hasps are securely fastened to the magazine and to the door frame.

These requirements do not apply to magazine doors that are adequately secured on the inside by means of a bolt, lock, or bar that cannot be actuated from the outside.

[T.D. ATF–87, 46 FR 40384, Aug. 7, 1981, as amended by T.D. ATF–298, 55 FR 21863, May 30, 1990]

§ 555.212 Smoking and open flames.

Smoking, matches, open flames, and spark producing devices are not permitted:

(a) In any magazine;

(b) Within 50 feet of any outdoor magazine; or

(c) Within any room containing an indoor magazine.

§ 555.213 Quantity and storage restrictions.

(a) Explosive materials in excess of 300,000 pounds or detonators in excess of 20 million are not to be stored in one magazine unless approved by the Director.

(b) Detonators are not to be stored in the same magazine with other explosive materials, except under the following circumstances:

(1) In a type 4 magazine, detonators that will not mass detonate may be stored with electric squibs, safety fuse, igniters, and igniter cord.

(2) In a type 1 or type 2 magazine, detonators may be stored with delay devices and any of the items listed in paragraph (b)(1) of this section.

§ 555.214 Storage within types 1, 2, 3, and 4 magazines.

(a) Explosive materials within a magazine are not to be placed directly against interior walls and must be stored so as not to interfere with ventilation. To prevent contact of stored explosive materials with walls, a nonsparking lattice work or other nonsparking material may be used.

(b) Containers of explosive materials are to be stored so that marks are visible. Stocks of explosive materials are to be stored so they can be easily counted and checked upon inspection.

(c) Except with respect to fiberboard or other nonmetal containers, containers of explosive materials are not to be unpacked or repacked inside a magazine or within 50 feet of a magazine, and must not be unpacked or repacked close to other explosive materials. Containers of explosive materials must be closed while being stored.

(d) Tools used for opening or closing containers of explosive materials are to be of nonsparking materials, except that metal slitters may be used for opening fiberboard containers. A wood wedge and a fiber, rubber, or wooden mallet are to be used for opening or closing wood containers of explosive materials. Metal tools other than nonsparking transfer conveyors are not to be stored in any magazine containing high explosives.

§ 555.215 Housekeeping.

Magazines are to be kept clean, dry, and free of grit, paper, empty packages and containers, and rubbish. Floors are to be regularly swept. Brooms and other utensils used in the cleaning and maintenance of magazines must have no spark-producing metal parts, and may be kept in magazines. Floors stained by leakage from explosive materials are to be cleaned according to instructions of the explosives manufacturer. When any explosive material has deteriorated it is to be destroyed in accordance with the advice or instructions of the manufacturer. The area surrounding magazines is to be kept clear of rubbish, brush, dry grass, or trees (except live trees more than 10 feet tall), for not less than 25 feet in all directions. Volatile materials are to be kept a distance of not less than 50 feet from outdoor magazines. Living foliage which is used to stabilize the earthen covering of a magazine need not be removed.

§ 555.216 Repair of magazines.

Before repairing the interior of magazines, all explosive materials are to be removed and the interior cleaned. Before repairing the exterior of magazines, all explosive materials must be removed if there exists any possibility that repairs may produce sparks or flame. Explosive materials removed from magazines under repair must be

(a) placed in other magazines appropriate for the storage of those explosive materials under this subpart, or

(b) placed a safe distance from the magazines under repair where they are to be properly guarded and protected until the repairs have been completed.

§ 555.217 Lighting.

(a) Battery-activated safety lights or battery-activated safety lanterns may be used in explosives storage magazines.

(b) Electric lighting used in any explosives storage magazine must meet the standards prescribed by the "National Electrical Code," (National Fire Protection Association, NFPA 70–81), for the conditions present in the magazine at any time. All electrical switches are to be located outside of the magazine and also meet the standards prescribed by the National Electrical Code.

(c) Copies of invoices, work orders or similar documents which indicate the lighting complies with the National Electrical Code must be available for inspection by ATF officers.

§ 555.218 Table of distances for storage of explosive materials.

Quantity of Explosives		Distances in feet							
Pounds over	Pounds not over	Inhabited buildings		Public highways with traffic volume 3000 or fewer vehicles/day		Passenger railways-public highways with traffic volume more than 3,000 vehicles/day		Separation of magazines	
		Barricaded	Unbarricaded	Barricaded	Unbarricaded	Barricaded	Unbarricaded	Barricaded	Unbarricaded
0	5	70	140	30	60	51	102	6	12
5	10	90	180	35	70	64	128	8	16
10	20	110	220	45	90	81	162	10	20
20	30	125	250	50	100	93	186	11	22
30	40	140	280	55	110	103	206	12	24
40	50	150	300	60	120	110	220	14	28
50	75	170	340	70	140	127	254	15	30
75	100	190	380	75	150	139	278	16	32
100	125	200	400	80	160	150	300	18	36
125	150	215	430	85	170	159	318	19	38
150	200	235	470	95	190	175	350	21	42
200	250	255	510	105	210	189	378	23	46
250	300	270	540	110	220	201	402	24	48
300	400	295	590	120	240	221	442	27	54
400	500	320	640	130	260	238	476	29	58
500	600	340	680	135	270	253	506	31	62
600	700	355	710	145	290	266	532	32	64
700	800	375	750	150	300	278	556	33	66
800	900	390	780	155	310	289	578	35	70
900	1,000	400	800	160	320	300	600	36	72
1,000	1,200	425	850	165	330	318	636	39	78
1,200	1,400	450	900	170	340	336	672	41	82
1,400	1,600	470	940	175	350	351	702	43	86
1,600	1,800	490	980	180	360	366	732	44	88
1,800	2,000	505	1,010	185	370	378	756	45	90
2,000	2,500	545	1,090	190	380	408	816	49	98
2,500	3,000	580	1,160	195	390	432	864	52	104
3,000	4,000	635	1,270	210	420	474	948	58	116
4,000	5,000	685	1,370	225	450	513	1,026	61	122
5,000	6,000	730	1,460	235	470	546	1,092	65	130
6,000	7,000	770	1,540	245	490	573	1,146	68	136
7,000	8,000	800	1,600	250	500	600	1,200	72	144
8,000	9,000	835	1,670	255	510	624	1,248	75	150
9,000	10,000	865	1,730	260	520	645	1,290	78	156
10,000	12,000	875	1,750	270	540	687	1,374	82	164
12,000	14,000	885	1,770	275	550	723	1,446	87	174
14,000	16,000	900	1,800	280	560	756	1,512	90	180
16,000	18,000	940	1,880	285	570	786	1,572	94	188
18,00	20,000	975	1,950	290	580	813	1,626	98	196
20,000	25,000	1,055	2,000	315	630	876	1,752	105	210
25,000	30,000	1,130	2,000	340	680	933	1,866	112	224
30,000	35,000	1,205	2,000	360	720	981	1,962	119	238
35,000	40,000	1,275	2,000	380	760	1,026	2,000	124	248
40,000	45,000	1,340	2,000	400	800	1,068	2,000	129	258
45,000	50,000	1,400	2,000	420	840	1,104	2,000	135	270
50,000	55,000	1,460	2,000	440	880	1,140	2,000	140	280
55,000	60,000	1,515	2,000	455	910	1,173	2,000	145	290
60,000	65,000	1,565	2,000	470	940	1,206	2,000	150	300
65,000	70,000	1,610	2,000	485	970	1,236	2,000	155	310
70,000	75,000	1,655	2,000	500	1,000	1,263	2,000	160	320
75,000	80,000	1,695	2,000	510	1,020	1,293	2,000	165	330
80,000	85,000	1,730	2,000	520	1,040	1,317	2,000	170	340
85,000	90,000	1,760	2,000	530	1,060	1,344	2,000	175	350
90,000	95,000	1,790	2,000	540	1,080	1,368	2,000	180	360
95,000	100,000	1,815	2,000	545	1,090	1,392	2,000	185	370
100,000	110,000	1,835	2,000	550	1,100	1,437	2,000	195	390
110,000	120,000	1,855	2,000	555	1,110	1,479	2,000	205	410
120,000	130,000	1,875	2,000	560	1,120	1,521	2,000	215	430
130,000	140,000	1,890	2,000	565	1,130	1,557	2,000	225	450
140,000	150,000	1,900	2,000	570	1,140	1,593	2,000	235	470
150,000	160,000	1,935	2,000	580	1,160	1,629	2,000	245	490
160,000	170,000	1,965	2,000	590	1,180	1,662	2,000	255	510
170,000	180,000	1,990	2,000	600	1,200	1,695	2,000	265	530
180,000	190,000	2,010	2,010	605	1,210	1,725	2,000	275	550
190,000	200,000	2,030	2,030	610	1,220	1,755	2,000	285	570
200,000	210,000	2,055	2,055	620	1,240	1,782	2,000	295	590
210,000	230,000	2,100	2,100	635	1,270	1,836	2,000	315	630
230,000	250,000	2,155	2,155	650	1,300	1,890	2,000	335	670
250,000	275,000	2,215	2,215	670	1,340	1,950	2,000	360	720
275,000	300,000	2,275	2,275	690	1,380	2,000	2,000	385	770

Table: American Table of Distances for Storage of Explosives (December 1910), as Revised and Approved by the Institute of Makers of Explosives—July, 1991.

Notes to the Table of Distances for Storage of Explosives

(1) Terms found in the table of distances for storage of explosive materials are defined in §555.11.

(2) When two or more storage magazines are located on the same property, each magazine must comply with the minimum distances specified from inhabited buildings, railways, and highways, and, in addition, they should be separated from each other by not less than the distances shown for "Separation of Magazines," except that the quantity of explosives contained in cap magazines shall govern in regard to the spacing of said cap magazines from magazines containing other explosives. If any two or more magazines are separated from each other by less than the specified "Separation of Magazines" distances, then such two or more magazines, as a group, must be considered as one magazine, and the total quantity of explosives stored in such group must be treated as if stored in a single magazine located on the site of any magazine of the group, and must comply with the minimum of distances specified from other magazines, inhabited buildings, railways, and highways.

(3) All types of blasting caps in strengths through No. 8 cap should be rated at 1½ lbs. (1.5 lbs.) of explosives per 1,000 caps. For strengths higher than No. 8 cap, consult the manufacturer.

(4) For quantity and distance purposes, detonating cord of 50 or 60 grains per foot should be calculated as equivalent to 9 lbs. of high explosives per 1,000 feet. Heavier or lighter core loads should be rated proportionately.

[T.D. ATF–87, 46 FR 40384, Aug. 7, 1981, as amended by T.D. ATF–400, 63 FR 45003, Aug. 24, 1998; T.D. ATF–446, 66 FR 16602, Mar. 27, 2001; T.D. ATF–446a, 66 FR 19089, Apr. 13, 2001]

§ 555.219 Table of distances for storage of low explosives.

Pounds		From inhabited building distance (feet)	From public railroad and highway distance (feet)	From above ground magazine (feet)
Over	Not over			
0	1,000	75	75	50
1,000	5,000	115	115	75
5,000	10,000	150	150	100
10,000	20,000	190	190	125
20,000	30,000	215	215	145
30,000	40,000	235	235	155
40,000	50,000	250	250	165
50,000	60,000	260	260	175
60,000	70,000	270	270	185
70,000	80,000	280	280	190
80,000	90,000	295	295	195
90,000	100,000	300	300	200
100,000	200,000	375	375	250
200,000	300,000	450	450	300

Table: Department of Defense Ammunition and Explosives Standards, Table 5–4.1 Extract; 4145.27 M, March 1969

§ 555.220 Table of separation distances of ammonium nitrate and blasting agents from explosives or blasting agents.

Donor weight (pounds)		Minimum separation distance of acceptor from donor when barricaded (feet)		Minimum thickness of artificial barricades (inches)
Over	Not over	Ammonium nitrate	Blasting agent	
0	100	3	11	12
100	300	4	14	12
300	600	5	18	12
600	1,000	6	22	12
1,000	1,600	7	25	12
1,600	2,000	8	29	12
2,000	3,000	9	32	15
3,000	4,000	10	36	15
4,000	6,000	11	40	15
6,000	8,000	12	43	20
8,000	10,000	13	47	20
10,000	12,000	14	50	20
12,000	16,000	15	54	25
16,000	20,000	16	58	25
20,000	25,000	18	65	25
25,000	30,000	19	68	30
30,000	35,000	20	72	30
35,000	40,000	21	76	30
40,000	45,000	22	79	35
45,000	50,000	23	83	35
50,000	55,000	24	86	35
55,000	60,000	25	90	35
60,000	70,000	26	94	40
70,000	80,000	28	101	40
80,000	90,000	30	108	40
90,000	100,000	32	115	40
100,000	120,000	34	122	50
120,000	140,000	37	133	50
140,000	160,000	40	144	50
160,000	180,000	44	158	50
180,000	200,000	48	173	50
200,000	220,000	52	187	60
220,000	250,000	56	202	60
250,000	275,000	60	216	60
275,000	300,000	64	230	60

Table: National Fire Protection Association (NFPA) Official Standard No. 492, 1968

Notes of Table of Separation Distances of Ammonium Nitrate and Blasting Agents from Explosives or Blasting Agents

(1) This table specifies separation distances to prevent explosion of ammonium nitrate and ammonium nitrate-based blasting agents by propagation from nearby stores of high explosives or blasting agents referred to in the table as the "donor." Ammonium nitrate, by itself, is not considered to be a donor when applying this table. Ammonium nitrate, ammonium nitrate-fuel oil or combinations thereof are acceptors. If stores of ammonium nitrate are located within the sympathetic detonation distance of explosives or blasting agents, one-half the mass of the ammonium nitrate is to be included in the mass of the donor.

(2) When the ammonium nitrate and/or blasting agent is not barricaded, the distances shown in the table must be multiplied by six. These distances allow for the possibility of high velocity metal fragments from mixers, hoppers, truck bodies, sheet metal structures, metal containers, and the like which may enclose the "donor." Where explosives storage is in bullet-resistant magazines or where the storage is protected by a bullet-resistant wall, distances and barricade thicknesses in excess of those prescribed in the table in §555.218 are not required.

(3) These distances apply to ammonium nitrate that passes the insensitivity test prescribed in the definition of ammonium nitrate fertilizer issued by the Fertilizer Institute. 1 Ammonium nitrate failing to pass the test must be stored at separation distances in accordance with the table in §555.218.

[1] Definition and Test Procedures for Ammonium Nitrate Fertilizer, Fertilizer Institute 1015–18th St. N.W. Washington, DC 20036.

(4) These distances apply to blasting agents which pass the insensitivity test prescribed in regulations of the U.S. Department of Transportation (49 CFR part 173).

(5) Earth or sand dikes, or enclosures filled with the prescribed minimum thickness of earth or sand are acceptable artificial barricades. Natural barricades, such as hills or timber of sufficient density that the surrounding exposures which require protection cannot be seen from the "donor" when the trees are bare of leaves, are also acceptable.

(6) For determining the distances to be maintained from inhabited buildings, passenger railways, and public highways, use the table in §555.218.

§ 555.221 Requirements for display fireworks, pyrotechnic compositions, and explosive materials used in assembling fireworks or articles pyrotechnic.

(a) Display fireworks, pyrotechnic compositions, and explosive materials used to assemble fireworks and articles pyrotechnic shall be stored at all times as required by this Subpart unless they are in the process of manufacture, assembly, packaging, or are being transported.

(b) No more than 500 pounds (227 kg) of pyrotechnic compositions or explosive materials are permitted at one time in any fireworks mixing building, any building or area in which the pyrotechnic compositions or explosive materials are pressed or otherwise prepared for finishing or assembly, or any finishing or assembly building. All pyrotechnic compositions or explosive materials not in immediate use will be stored in covered, non-ferrous containers.

(c) The maximum quantity of flash powder permitted in any fireworks process building is 10 pounds (4.5 kg).

(d) All dry explosive powders and mixtures, partially assembled display fireworks, and finished display fireworks shall be removed from fireworks process buildings at the conclusion of a day's operations and placed in approved magazines.

[T.D. ATF–293, 55 FR 3722, Feb. 5, 1990, as amended by T.D. ATF–400, 63 FR 45004, Aug. 24, 1998]

§ 555.222 Table of distances between fireworks process buildings and between fireworks process and fireworks nonprocess buildings.

Net weight of fireworks1 (pounds)	Display fireworks2 (feet)	Consumer fireworks3 (feet)
0-100	57	37
101-200	69	37
201-300	77	37
301-400	85	37
401-500	91	37
Above 500	Not permitted4 5	Not permitted4 5

[1] Net weight is the weight of all pyrotechnic compositions, and explosive materials and fuse only.

[2] The distances in this column apply only with natural or artificial barricades. If such barricades are not used, the distances must be doubled.

[3] While consumer fireworks or articles pyrotechnic in a finished state are not subject to regulation, explosive materials used to manufacture or assemble such fireworks or articles are subject to regulation. Thus, fireworks process buildings where consumer fireworks or articles pyrotechnic are being processed shall meet these requirements.

[4] A maximum of 500 pounds of in-process pyrotechnic compositions, either loose or in partially-assembled fireworks, is permitted in any fireworks process building. Finished display fireworks may not be stored in a fireworks process building.

[5] A maximum of 10 pounds of flash powder, either in loose form or in assembled units, is permitted in any fireworks process building. Quantities in excess of 10 pounds must be kept in an approved magazine.

[T.D. ATF–293, 55 FR 3723, Feb. 5, 1990, as amended by T.D. ATF–400, 63 FR 45004, Aug. 24, 1998]

§ 555.223 Table of distances between fireworks process buildings and other specified areas.

Net weight of fireworks1 (pounds)	Display fireworks1 (feet)	Consumer fireworks2 (feet)
0-100	200	25
101-200	200	50
201-300	200	50
301-400	200	50
401-500	200	50
Above 500	Not permitted	Not permitted

Distance from Passenger Railways, Public Highways, Fireworks Plant Buildings used to Store Consumer Fireworks and Articles Pyrotechnic, Magazines and Fireworks Shipping Buildings, and Inhabited Buildings. [3 4 5]

[1] Net weight is the weight of all pyrotechnic compositions, and explosive materials and fuse only.

[2] While consumer fireworks or articles pyrotechnic in a finished state are not subject to regulation, explosive materials used to manufacture or assemble such fireworks or articles are subject to regulation. Thus, fireworks process buildings where consumer fireworks or articles pyrotechnic are being processed shall meet these requirements.

[3] This table does not apply to the separation distances between fireworks process buildings (see § 555.222) and between magazines (see §§ 555.218 and 555.224).

[4] The distances in this table apply with or without artificial or natural barricades or screen barricades. However, the use of barricades is highly recommended.

[5] No work of any kind, except to place or move items other than explosive materials from storage, shall be conducted in any building designated as a warehouse. A fireworks plant warehouse is not subject to § 555.222 or this section, tables of distances.

[T.D. ATF–293, 55 FR 3723, Feb. 5, 1990, as amended by T.D. ATF–400, 63 FR 45004, Aug. 24, 1998]

§ 555.224 Table of distances for the storage of display fireworks (except bulk salutes).

Net weight of fireworks1 (pounds)	Distance between magazine and inhabited building, passenger railway, or public highway3 4 (feet)	Distance between magazines2 3 (feet)
0-1000	150	100
1,001-5,000	230	150
5,001-10,000	300	200
Above 10,000	Use Table § 55.218	

[1] Net weight is the weight of all pyrotechnic compositions, and explosive materials and fuse only.

[2] For the purposes of applying this table, the term "magazine" also includes fireworks shipping buildings for display fireworks.

[3] For fireworks storage magazines in use prior to (30 days from the date of publication of the final rule in the Federal Register), the distances in this table may be halved if properly barricaded between the magazine and potential receptor sites.

[4] This table does not apply to the storage of bulk salutes. Use table at § 555.218.

[T.D. ATF–293, 55 FR 3723, Feb. 5, 1990, as amended by T.D. ATF–400, 63 FR 45004, Aug. 24, 1998]

Questions and Answers (Revised 1/07)
18 U.S.C. Chapter 40 and 27 CFR Part 555

Introduction

The following list of Questions and Answers are intended to aid you in gaining a better understanding of:

18 U.S.C. Chapter 40 – Importation, Manufacture, Distribution and Storage of Explosive Materials and the implementing regulations issued within:

27 CFR Part 555 – Commerce in Explosives

This listing is not all-inclusive. However it contains some of the most frequently asked questions that ATF receives. These questions and answers are intended only as a general overview. To determine how the law and regulations apply to your specific circumstances, you must refer directly to the applicable law and regulation or contact your local ATF Office of Industry Operations. Also, please be aware that both the law and regulations are subject to change. Please contact your local ATF office for the most up-to-date information. You can also find the latest Federal explosives regulations on the ATF website: www.atf.gov.

Unless otherwise stated, these Questions and Answers apply only to Federal law and regulations. States and local jurisdictions have, in many cases, enacted their own requirements relating to explosives. Check with appropriate State or local authorities for information on those requirements. Compliance with Federal law and regulations does not exempt any person from compliance with any State or local requirements.

A Table of Contents and a Subject Index has been included for your convenience. The Index is located at the end of the Questions and Answers section.

Table of Contents

General Questions

1. Who is affected by the Federal explosives law?

The law affects all persons who import, manufacture, deal in, purchase, use, store, or possess explosive materials. It also affects those who ship, transport or cause to be transported, or receive explosive materials. Also, see 18 U.S.C. 845 and 27 CFR 555.141 for exemptions.

2. What changes were brought about by the Safe Explosives Act?

Among other things, the Act mandated that all persons who wish to receive or transport explosive materials must first obtain a Federal explosives license or permit. In addition, the act imposed new restrictions on who may lawfully receive and possess explosive materials (See question 4). All Federal explosive licensees and permittees and their responsible persons and employees authorized to possess explosives are affected by the new requirements and background checks mandated by the Act.

3. Can I obtain general information from ATF on the Internet?

Yes. ATF maintains a website on the Internet at www.atf.gov.

4. Does the law make some classes of persons ineligible to receive a Federal license to import, manufacture, or deal in explosive materials or to receive a Federal explosives permit?

Yes. A license or permit will not be issued to any person who:

(a) Is under indictment for, or who has been convicted in any court, of a crime punishable by imprisonment for a term exceeding one year;

(b) Is a fugitive from justice;

(c) Is an unlawful user of or addicted to any controlled substance (as defined in section 102 of the Controlled Substances Act (21 U.S.C. 802));

(d) Has been adjudicated a mental defective or who has been committed to a mental institution;

(e) Is an alien (with certain exceptions);

(f) Has been discharged from the armed forces under dishonorable conditions; or,

(g) Having been a citizen of the United States, has renounced his citizenship. [18 U.S.C. 842(i), 843(b)(1); 27 CFR 555.49(b)(2)(i)]

5. Are there classes of persons to whom the distribution of explosive materials by licensees is prohibited?

Yes. No person shall knowingly distribute explosive materials to any individual listed in Question 4, or to an individual who is under 21 years of age. [18 U.S.C. 842(d); 27 CFR 555.26(d)]

6. What other distributions of explosive materials by licensees and permittees are prohibited?

A licensee or permittee shall not knowingly distribute any explosive materials to any person who:

(a) Is not a licensee [18 U.S.C. 842(b), 27 CFR 555.105, 555.106];

(b) Is not a holder of a user permit [18 U.S.C. 842(b); 27 CFR 555.105, 555.106];

(c) Is not a holder of a limited permit who resides in the same State where distribution is made and in which premises of the transferor are located. [18 U.S.C. 842(b); 27 CFR 555.105, 555.106];

(d) The licensee has reason to believe intends to transport such explosive materials into a State where the purchase, possession, or use of explosive materials is prohibited or which does not permit its residents to transport or ship explosive materials into the State or to receive explosive materials in the State. [18 U.S.C. 842(c); 27 CFR 555.106]

(e) Is in any State where the purchase, possession, or use by such person of such explosive materials would be in violation of any State law or any published ordinance applicable at the place of distribution. [18 U.S.C. 842(e); 27 CFR 555.106(b)(2)]

7. Does Federal law prohibit certain persons from receiving or possessing explosive materials?

Yes. The law prohibits the receipt or possession of explosive materials by any person listed in question 4. [18 U.S.C. 842(i); 27 CFR 555.26, 555.49(b)]

8. May a licensed manufacturer, importer, or dealer distribute explosive materials to nonlicensees and nonpermittees?

No. Every person who receives explosive materials must first obtain a Federal explosives license or permit. Distribution of explosive materials to persons who do not hold a license or permit is unlawful. Also, see 18 U.S.C. 845 and 27 CFR 555.141 for exemptions. [18 U.S.C. 842(a), 842(b); 27 CFR 555.26(a), 555.106]

9. Does Federal law provide penalties for purchasers who give false information at the time of purchasing explosive materials?

Yes. The penalty for knowingly providing false information or misrepresented identification is a maximum 10 years' imprisonment and/or a fine not exceeding $250,000. [18 U.S.C. 842(a)(2), 844(a)]

10. Is the theft of explosive materials, as well as the possession of stolen explosive materials, a Federal crime?

Yes. It is a Federal crime for any person to steal any explosive materials. It is also a Federal crime for any person to receive, possess, transport, ship, conceal, store, barter, sell, dispose of, or pledge or accept as security for a loan any stolen explosive materials. [18 U.S.C. 842(h), 844(k), (l)]

11. Are thefts and losses of explosive materials required to be reported to ATF?

Yes. Any licensee or permittee who has knowledge of the theft or loss of any explosive materials from his or her stock shall, within 24 hours of discovery, report the theft or loss by telephoning 800-461-8841 (Monday-Friday 8:00a.m.– 5:00 p.m. Eastern Time) or 888-283-2662 (after hours and weekends) (nationwide toll free number) and on ATF Form 5400.5, "Report of Theft or Loss – Explosive Materials", in accordance with the instructions on the form. The theft or loss shall also be reported to appropriate local authorities. The same requirements are imposed upon persons other than licensees and permittees, except that nonlicensees and nonpermittees, other than carriers, need not report a theft or loss on Form 5400.5, but must report the theft or loss by telephone, using the same numbers: 800-461-8841 (Monday-Friday 8:00a.m.–5:00 p.m. Eastern Time) or 888-283-2662 (after hours and weekends) and in writing to the nearest ATF office. The theft or loss shall also be reported to appropriate local authorities. Carriers of explosive materials must report a theft or loss by telephone but need not make the report on the ATF form or in writing. See 27 CFR 555.30 for the specific information required to be reported in connection with a theft or loss. [18 U.S.C. 842(k), 18 U.S.C. 844(p); 27 CFR 555.30]

12. My company holds a Federal explosives license and after conducting an inventory of our explosives on hand, we noticed one case of dynamite missing. After double-checking all Daily Summaries of Magazine Transactions, invoices, and delivery sheets, we still cannot reconcile the discrepancy. What are we required to do?

This should be considered a theft or loss of explosive materials. As stated in the answer to Question 11, you must report the theft or loss of explosive materials, within 24 hours of discovery, to ATF by telephone (toll free: 800-461-8841 (Monday-Friday 8:00 a.m.– 5:00 p.m. Eastern Time) or 888-283-2662 (after hours and weekends). ATF Form 5400.5, "Report of Theft or Loss – Explosive Materials", must then be completed and forwarded in accordance with the instructions on the form. [18 U.S.C. 842(k); 27 CFR 555.30]

13. May ATF conduct warrantless inspections of licensees' and permittees' records of explosives materials, stocks of such materials, and magazines?

Except for limited permit holders, any ATF officer may, without a warrant, enter during business hours the premises, including places of storage, of any licensee or permittee for the purpose of inspecting or examining any records or documents required to be kept by the law and regulations and any explosive materials kept or stored at the premises. For inspection purposes, "business hours" includes hours during which business is actually conducted, not just those hours stated on license applications. Any licensee or permittee who refuses to permit the inspection or examination is subject to having his or her license or permit revoked, as well as to denial of an application to renew the license or permit. For limited permit holders, an ATF officer may inspect the places of storage for explosive materials of either an applicant for a limited permit or at the time of renewal of such permit, but in no event shall such inspection occur more than once every three years. [18 U.S.C. 843(b)(4), 18 U.S.C. 843(f)]

14. Will ATF investigate accidents involving explosive materials?

ATF is authorized to inspect the site of any accident or fire where there is reason to believe that explosive materials were involved. Other Federal agencies, or State or local agencies, may also investigate such incidents, depending on the circumstances. [18 U.S.C. 846(a); 27 CFR 555.31]

15. Is black powder subject to regulation under Federal explosives laws?

Black powder is an explosive material for purposes of Federal explosives laws and regulations. However, the law exempts from regulation commercially manufactured black powder in quantities not exceeding 50 pounds (as well as percussion caps, safety and pyrotechnic fuses, quills, quick and slow matches, and friction primers) intended to be used solely for sporting, recreational, or cultural purposes in antique firearms as defined in 18 U.S.C. 921(a)(16) or in antique devices exempted from the term "destructive device" in 18 U.S.C. 921(a)(4). However, persons engaged in the business of importing, manufacturing, or dealing in black powder in any quantity must have a Federal explosives license. [18 U.S.C. 841(c), 841(d), 845(a)(5); 27 CFR 555.11: definitions of "explosives" and "explosive materials", 555.141(b)]

16. Is small arms ammunition subject to regulation under Federal explosives laws?

No. The law specifically exempts small arms ammunition and components thereof. (See also Question 81.) [18 U.S.C. 845(a)(4)]

17. Are binary explosives subject to regulation under Federal explosives laws?

Until the compounds are mixed, they are not classified as explosives and, therefore, are not subject to control. However, once mixed, binary explosives are "explosive materials" and are subject to all applicable Federal requirements. A person who mixes or combines compounds of binary explosives for the purpose of sale or distribution or for the person's own business use is a "manufacturer" of explosive materials and must be licensed as a manufacturer under the law. [18 U.S.C. 841(h); 27 CFR 555.11: definition of "manufacturer"]

18. Does ATF have any regulations governing the actual transportation of explosive materials?

Federal explosives laws and regulations generally prohibit any person from transporting explosive materials interstate or intrastate unless the person has a Federal explosives license or permit. Also, the transportation of stolen explosives materials is a Federal crime (see also Question 10). However, the law exempts from regulation under 18 U.S.C. Chapter 40 and 27 CFR Part 555 aspects of the transportation of explosive materials via railroad, water, highway, or air which are regulated by the United States Department of Transportation, and the Department of Homeland Security, and agencies thereof and which pertain to safety and security. [18 U.S.C. 842(a)(3), 842(h), 845(a)(1); 27 CFR 555.26, 555.28, 555.141(a)(1)]

19. Are common or contract carriers required to obtain a Federal explosives license or permit to transport explosive materials?

No. The actual transportation of explosive materials by carriers is subject to Department of Transportation or Department of Homeland Security regulations. [18 U.S.C. 845(a)(1); 27 CFR 555.141(a)(1)]

20. What is the "Explosives List"?

The Explosives List is a comprehensive (but not all-inclusive) listing of explosive materials which have been determined to be within the coverage of Chapter 40. The list is published annually by ATF (the most recent list can be found under the Explosives Laws and Regulations link on the ATF website). [18 U.S.C. 841(d); 27 CFR 555.23]

21. May a person under the age of 21 be lawfully employed by an explosives business and lawfully receive, possess, and use explosive materials on behalf of the business?

Yes. Federal explosives law prohibits any person from distributing explosive materials to persons under 21 years of age. However, it does not prohibit the delivery to or possession of explosive materials by persons under the age of 21 who are receiving or using the materials on behalf of their employers to whom the materials were lawfully sold. [18 U.S.C. 842(d), (i); 27 CFR 555.11 (definition of "distribute"), 555.26, 555.106(b)(1)]

22. ATF regulations require explosive materials to be stored at certain minimum distances from a "public highway". What is a "public highway" for purposes of the regulations?

The term "highway" is defined in 27 CFR 555.11 as "any public street, public alley, or public road, including a privately financed, constructed, or maintained road that is regularly and openly traveled by the general public." Privately financed, constructed, or maintained roads that are marked and barricaded in a manner that prevents access by the general public do not fall within the meaning of the term and would, therefore, be exempt from table of distance requirements. [27 CFR 555.11: definition of "highway"](See also ATF Ruling 2005-2)

23. Is an airport runway or taxiway considered a public highway for purposes of the Table of Distances for storage of explosive materials?

No. However, airport terminals are considered inhabited buildings for Table of Distance requirements.

24. How is shock tube regulated by ATF?

Shock tube contains highly explosive material. However, it may be stored as a low explosive when not attached to a detonator. [27 CFR 555.202(b), 555.213]

25. What is an EX number?

An EX number is a number, preceded by the prefix "EX-", which is issued and used by the Department of Transportation (DOT) to identify an explosive which has been tested and classified by DOT. See U.S. Department of Transportation regulations at 49 CFR 171.8 and 49 CFR 173.56.

26. What is a UN number?

A UN (United Nations) number is used by DOT as a method of identification and classification of products for shipping purposes. UN numbers are different from the hazard class or division designations used by DOT (for example, 1.1, 1.2, 1.3, 1.4, and 1.5). ATF regulations in 27 CFR Part 555 also use UN numbers to help identify certain explosives. [27 CFR 555.11]

27. Can Federal explosives disabilities resulting from a conviction of a crime punishable by imprisonment for a term exceeding one year be removed if the conviction is expunged or set aside or the convicted person has received a pardon for the offense or has had his or her civil rights restored?

A person convicted of, or under indictment for, a "crime punishable by imprisonment for a term exceeding one year" may not lawfully receive or possess explosive materials or be issued a Federal explosives license or permit. The term "crime punishable by imprisonment for a term exceeding one year" does not include offenses pertaining to antitrust violations, unfair trade practices, restraints of trade, or any State offense (other than one involving a firearm or explosive) classified as a misdemeanor and punishable by imprisonment for 2 years or less. There are only 3 means by which Federal explosives disabilities resulting from a conviction of, or indictment for, a "crime punishable by imprisonment for a term exceeding one year" can be removed:

(a) A decision of a court invalidating a conviction on the basis that the conviction was unconstitutional;

(b) In the case of a Federal conviction, a presidential pardon; and

(c) The granting of relief from Federal explosives disabilities by ATF pursuant to the filing of a relief application with the Director. Information on how to apply for relief and contact information for the Relief of Disabilities Section is available on the ATF website. [18 U.S.C. 841(l), 842(d),(i), 845(b); 27 CFR 555.11: definition of "crime punishable by imprisonment for a term exceeding one year", 555.26(c), 555.142]

Licenses and Permits

28. Who needs a Federal explosives license or permit?

All persons who wish to transport, ship, cause to be transported, or receive explosive materials must first obtain a Federal explosives license or permit. Certain exemptions apply. [18 U.S.C. 842(b); 18 U.S.C. 845; 27 CFR 555.26(a), 27 CFR 555.141]

29. Who is eligible for a Federal explosives license or permit?

The Chief, Federal Explosives Licensing Center, will approve a properly completed application for a license or permit on ATF Form 5400.13/5400.16 if the applicant:

(a) Is not a person prohibited from possessing or receiving explosive materials under 18 U.S.C. 842(i) and none of the applicant's "responsible persons" are prohibited under section 842(1); (see also Question 4 in General Q&A);

(b) Has not willfully violated any provision of Chapter 40 or the regulations in 27 CFR Part 555;

(c) Has not knowingly withheld information or has not made any false or fictitious statement intended or likely to deceive, in connection with the application,

(d) Has premises in a State from which he intends to conduct business or operations;

(e) Has storage for the class (as described in 27 CFR 555.202) of explosive materials described on the application;

(f) Has certified in writing that he is familiar with and understands all published State laws and local ordinances relating to explosive materials for the location in which he intends to do business;

(g) Has submitted the certificate required by section 21 of the Federal Water Pollution Control Act, as amended (33 U.S.C. 1341) [18 U.S.C. 843(b); 27 CFR 555.49(b)];

(h) None of the applicant's employees authorized to possess explosives are prohibited persons under 18 U.S.C. 842(i); and

(i) In the case of an applicant for a limited permit, the applicant has certified that the applicant will not receive explosive materials on more than 6 occasions during the 12-month period for which the limited permit is valid.

30. What activities are covered by licenses and permits?

Licenses allow persons to engage in the business of importing, manufacturing, or dealing in explosive materials. Any individual or business entity intending to engage in any of these activities must first obtain a license. A user permit allows the receipt and transportation of explosive materials. A limited permit allows the receipt of explosive materials from a licensee or permittee within the permittee's state of residence only, and on no more than six occasions in the 12-month period during which the limited permit is valid. A limited permit does not authorize the receipt or transportation of explosive materials in interstate or foreign commerce. [27 CFR 555.11: definitions of "importer", "manufacturer", "dealer", "limited permit", and "user permit", 555.41]

A separate license is needed for each business premises where an explosives business or activity is conducted. Only a single user permit is needed by a permittee who uses explosives in more than one location. [27 CFR 555.41]

31. What is a Limited Permit?

A limited permit is for persons who wish to transport, ship, cause to be transported, or receive explosive materials in intrastate commerce only. This permit is designed for the infrequent receipt of explosive materials by intrastate users. The limited permit will allow a purchaser to receive explosive materials on no more than six separate occasions from in-state licensees or permittees during the 12-month period of the permit. The limited permit does not allow the holder to transport, ship, cause to be transported, or receive explosive materials in interstate commerce.

32. What is the duration of a license or permit?

(a) A user license or permit is valid for a period of 3 years.

(b) The user-limited permit is valid only for a single purchase transaction.

(c) Limited permits are valid for no more than six separate receipts of explosive materials during a 12-month period. [27 CFR 555.51]

33. What are the fees for licenses and permits?

Each license applicant must pay a fee of $200 for obtaining a 3-year license, a separate license and fee being required for each business premises. The fee for renewal of a license is $100 for a 3-year license. [27 CFR 555.42]

Each applicant for a user permit must pay a fee of $100 for a 3-year permit, and each applicant for a user-limited permit (nonrenewable) must pay a fee of $75. The fee for renewal of a user permit is $50 for a 3-year permit. [27 CFR 555.43]

Each applicant for a limited permit must pay a fee of $25 for a 1-year limited permit. The fee for renewal of a limited permit is $12 for a 1-year limited permit. [27 CFR 555.43]

34. Will the Government investigate an application for a license or permit?

ATF will investigate any applicant before issuing a license or permit. Additionally, ATF must inspect places of storage and conduct background checks on responsible persons and employee possessors authorized to possess explosives [18 U.S.C. 843(b); 18 U.S.C. 843(h); 27 CFR 555.33, 27 CFR 555.49(b)]

35. What may a licensed explosives dealer do?

A licensed dealer may engage in the business of distributing explosive materials at wholesale or retail [27 CFR 555.11: definition of "dealer"]

36. What may a licensed explosives importer do?

A licensed importer may engage in the business of importing or bringing explosive materials into the United States for purposes of sale or distribution. It is not necessary for a licensed importer to also obtain a dealer's license to engage in business on his or her licensed premises as a dealer in explosive materials (see also Question 52 and 53) [27 CFR 555.11: definition of "importer", 27 CFR 555.41(b)(2)]

37. When is a manufacturer's license required?

A manufacturer's license is required by persons engaged in the business of manufacturing explosive materials for sale, distribution, or for their own business use. For example, persons engaged in the business of providing a blasting service using explosives of their own manufacture would be required to have a manufacturer's license. Persons who manufacture explosives for their personal, non-business use are not required to have a manufacturer's license. However, no person may ship, transport, cause to be transported, or receive explosive materials unless such person holds a license or permit. [27 CFR 555.11: definition of "manufacturer", 555.41(b)] A separate manufacturer's license is not required by a licensed manufacturer for the purpose of on-site manufacture, for example, mixing binary explosives or making blasting agents at a quarry or other job site. It is not necessary for a licensed manufacturer to also obtain a dealer's license to engage in business on his or her licensed premises as a dealer in explosive materials (see also Question 52 and 53) [27 CFR 555.11: definition of "manufacturer", 555.41(b)(2)]

38. How do I apply for a Federal explosives license or permit?

You can request an application for a Federal explosives license or permit from the Federal Explosives Licensing Center at 877-283-3352 or from the ATF Distribution Center at 703-455-7801. As part of the application process, you must complete and submit an ATF Form 5400.13/5400.16, Application for Explosives License or Permit. You must also submit the names, identifying information, fingerprints, and photographs of all responsible persons. In addition, you must submit the names and identifying information of all employees who are authorized to possess explosive materials in the course of their employment on ATF Form 5400.28, Employee Possessor Questionnaire. [27 CFR 555.45(c)]

39. Who is a "responsible person"?

Federal explosives laws define a "responsible person" as an individual who has the power to direct the management and policies of the applicant pertaining to explosive materials. Responsible persons generally include sole proprietors and explosives facility site managers. In the case of a corporation, association, or similar organization, responsible persons generally include only those corporate directors/officers, and stockholders, who have the power to direct management and policies as they pertain to explosive materials.

For example, a corporate vice president whose duties include acquiring and approving contracts with explosives distributors would be considered a responsible person. Other corporate officials whose duties do not include the power to direct the management and policies of the applicant pertaining to explosive materials, for example, a vice president responsible solely for human resources, would not typically be considered a responsible person. Each applicant for a license or permit must assess the corporate and other management responsibilities for all key personnel and determine whether or not these duties place the individual in the position of being a responsible person. [18 U.S.C. 841(s), 27 CFR 555.11: definition of "responsible person"]

40. Who is a "possessor of explosives"?

A possessor of explosives is any employee of a license or permit holder or any employee of an applicant for a license or permit who has or will have actual physical possession of explosive materials or who has or will have constructive possession of explosive materials. For example, persons who physically handle explosive materials would be considered to be actual possessors of explosive materials. This would include employees who directly handle explosive materials as part of the production process; employees who handle explosive materials in order to ship, transport, or sell them; and employees, such as blasters and their helpers who actually use explosive materials. A constructive possessor is any person who has access to explosive materials, without physically handling them. For example, a supervisor at a construction site who keeps keys for storage magazines in which explosives are stored or who directs the use of explosive materials by other employees has constructive possession of explosives.

41. Why is it necessary to provide new and additional information on responsible persons and employee possessors of explosives?

The law requires this information for ATF to conduct background checks on all responsible persons and employee possessors to restrict the availability of explosives to authorized persons only and to reduce the risk of prohibited persons acquiring explosive materials. [18 U.S.C. 843(h); 27 CFR 555.33, 555.45(c)]

42. When will I need to submit the identifying information for my responsible persons and employee possessors of explosives?

(a) All license and permit applicants and any renewal applicants must submit identifying information for responsible persons and employee possessors (and fingerprints and photographs for responsible persons) upon submission of an original or renewal application.

(b) Any new responsible person added after a license or permit has been issued by ATF must be reported to ATF within 30 days. However, the submission of fingerprints and photographs by the new responsible person is required only at the time of any subsequent renewal.

(c) For all licenses and permits (new and renewal), any new employee possessors must be reported to ATF within 30 days of hire on the Employee Possessor Questionnaire form (ATF F 5400.28). [27 CFR 555.45(c), 27 CFR 555.57(b)]

43. How do I get my fingerprints taken?

Fingerprints must be submitted on Fingerprint Identification Cards, FD-258 that have been issued by ATF. The fingerprint cards must contain the following ORI information: WVATF0900; ATF-NATL EXPL LIC, MARTINSBURG WV. These fingerprint cards may be obtained by contacting the Federal Explosives Licensing Center at 877-283-3352 or the ATF Distribution Center at 703-455-7801. The fingerprint cards must be completed by your local law enforcement authority.

44. Will ATF notify me whether or not my responsible persons and employee possessors have passed their background checks?

Yes. A "Notification of Clearance" will be issued directly to all license or permit holders advising whether their responsible persons and employee possessors have been cleared to possess explosive materials, or are or may be prohibited from possessing explosives. These notices must be retained as part of the license or permit holders permanent records. In addition, letters of clearance or denial will be issued directly to responsible persons and employee possessors. [27 CFR 555.33]

45. What notification will I receive if one of my responsible persons or employee possessors does not pass their ATF background check?

If an individual does not pass the background check, a letter will be sent to the licensee or permittee who submitted the individual's name indicating that the individual was denied. A letter will also be sent to that individual explaining the prohibition and outlining appeal and relief procedures, as may be applicable. Unless and until an appeal overturns the denial or relief from disabilities is granted, that individual may not lawfully possess explosives. [27 CFR 555.33]

46. Who will conduct the background checks on applicants, responsible persons, and possessors?

ATF will perform the background checks. If employers wish to require their own background checks as a condition of employment, they may do so. However, such a background check will not be accepted in place of the ATF background check. [27 CFR 555.33]

47. May I sell black powder without a license?

No. Anyone who engages in the business of selling black powder, regardless of quantity, must be licensed as an explosives dealer. [27 CFR 555.41(b)]

48. Is a manufacturer's license required to acquire and mix binary explosives?

If the individual purchasing the binary explosives is engaged in the business of manufacturing explosives, i.e., mixes and uses them in the operation of a commercial business (for example, operating a quarry, or providing the service of removing stumps or boulders from a farm field), then a manufacturer's license is required.

An individual farmer who merely wishes to mix the binary explosives to remove obstacles from his field and provides no other outside service would not need a manufacturer's license.

Please note, however: A Federal explosives license or permit would be required to obtain any explosive device, such as detonators, used to initiate the mixed binary explosives. In addition, transportation of any explosive material, including mixed binary explosives, without a Federal license or permit is prohibited. [27 CFR 555.11: Definition of "manufacturer"; 27 CFR 555.26, 555.41(b)]

49. What is theatrical flash powder and is there a license for its manufacture?

Theatrical flash powder is flash powder commercially manufactured in premeasured kits not exceeding 1 ounce in weight, and mixed immediately prior to use and intended for use in events such as theatrical shows, stage plays, band concerts, magic acts, thrill shows, and clown acts in circuses. A manufacturer's license allows on-site manufacturers to operate nationally on one license issued to their principal place of business. [27 CFR 555.11: definitions of "flash powder" and "theatrical flash powder", 555.41(b)]

50. Is a separate license required for each location where business is conducted?

Yes. A separate license is required for each location where business is conducted. However, a separate license is not required for:

(a) Facilities used only for the storage of explosive materials;

(b) Locations used solely for the storage of records relating to the business; and

(c) Licensed manufacturers' on-site manufacturing. [27 CFR 555.41(b)]

51. Must a person who engages in the business of both manufacturing and importing at the same location have both licenses?

Yes. The licenses for manufacturing and importing allow a person to engage in separate and distinct activities and a separate license is required for each activity. However, a manufacturer or an importer does not need a separate dealer's license to also distribute explosive materials from the licensed premises. [27 CFR 555.41(b)]

52. Does a licensed manufacturer, importer, or dealer need a permit to use explosive materials?

No. No licensee will be required to obtain a user permit to lawfully transport, ship, or receive explosive materials in interstate or foreign commerce. [27 CFR 555.41(b)(2)]

53. Does a Federal license or permit exempt the holder from State or local requirements?

No. A license or permit confers no right or privilege to conduct business or operations, including storage, contrary to State or other law. All legal requirements must be followed, whether Federal, State, or local. [18 U.S.C. 848; 27 CFR 555.62]

54. Who is authorized to import explosive materials?

Any licensed importer is authorized to engage in the business of importing explosive materials for sale, distribution, or their own use. Any licensed manufacturer, dealer, or holder of a user permit may import explosive materials for their own use only. Licensees and user permittees importing explosive materials must provide to the U.S. Customs and Border Protection (CBP) a copy of the license or permit. Note, however, that in the case of certain military explosives or propellant powder or other components of small arms ammunition, Federal firearms regulations require the importer to provide an approved ATF Form 6 to the CBP. [27 CFR 555.41(b)(2), 555.41(b)(3), 447.21, 555.108(a), 555.183, 478.113]

55. How may an employee of an explosives licensee or permittee qualify to accept delivery of explosive materials for the employer?

The employee must be on the current list of representatives or agents authorized to accept delivery of explosive materials on behalf of the employer and be an authorized employee possessor of explosives. [27 CFR 555.103(b), 555.105(b)]

56. When an explosives licensee or permittee sends one of their truck drivers to the distributor's premises to pick up explosive materials that have been purchased by the licensee or permittee, will the driver be required to sign any forms?

No, however the driver is required to furnish the seller with an identification document as defined in 27 CFR Part 555.11. [27 CFR 555.103(b), 555.105(b)]

57. Will a licensee or permittee be notified in advance when the license or user permit needs to be renewed?

Generally, prior to expiration of the license or permit, a licensee or permittee will be notified. The application form must be completed and filed with ATF before expiration of the current license or permit for the renewal to be considered timely. However, if a licensee or permittee does not receive a renewal notification, it is still that licensee's or permittee's responsibility to ensure that an application is filed prior to expiration of the current license or permit. [27 CFR 555.46]

58. I have timely filed my application for renewal of my license (or user's permit) but I have not received my new license (or permit). May I continue in business even though the expiration date shown on my license or permit has passed? If so, how long?

Yes. You may continue to operate the business pursuant to your current license or permit until the application for renewal is acted upon. [5 U.S.C. 558]

59. Can a license or permit be revoked?

Yes. The Director, Industry Operations for the ATF Field Division in which a licensee or permittee is located may revoke a license or permit if the holder has violated any provision of 18 U.S.C. Chapter 40 or its implementing regulations or has become ineligible to receive explosive materials under 18 U.S.C. 842(i). [18 U.S.C. 843(d); 27 CFR 555.71, 555.74]

60. If a Federal explosives licensee or permittee is indicted for or convicted of a "crime punishable by imprisonment for a term exceeding one year", may he or she continue operations under the license or permit?

As stated in the answer to Question 4 in General Q&A, a person under indictment for, or convicted of, a crime punishable by imprisonment for a term exceeding one year is not eligible to be issued a license or permit. However, a licensee or permittee who is indicted for, or convicted of, such a crime during the term of his or her existing license or permit is not barred from licensed or permit operations for 30 days after the date of the indictment or the date the conviction becomes final. If the licensee or permittee files an application for relief from disabilities within such 30-day period, he or she may continue licensed or permit operations while the application is pending. If a relief application is not filed during that period, the licensee or permittee may not continue operations beyond such 30-day period. The right of a licensee to continue licensed or permitted operations beyond such 30-day period is also conditioned on the licensee or permittee timely filing a license or permit renewal application disclosing that the applicant has been indicted for, or convicted of, the crime. A licensee or permittee may not continue operations beyond 30 days following the date the Director issues notification that the relief application has been denied. [18 U.S.C. 845(b); 27 CFR 555.142]

61. May a licensed dealer make a sale to a holder of a limited permit in an adjoining State?

No. Sales may not be made to limited permittees who are out-of-State residents. [18 U.S.C. 842(a); 27 CFR 555.11: definition of "limited permit", 555.41(b)(3)]

Recordkeeping

62. Does a licensee or permittee have to keep records of the acquisition, distribution, and storage of explosive materials?

Yes. Licensees and permittees must keep records of acquisitions, dispositions, and storage of explosive materials. [18 U.S.C. 842(f), 847; 27 CFR 555.107, 555.122-.125, and 555.127, Subpart G]

63. How do licensees and permittees account for explosive quantities in their records?

If acquisitions are recorded by weight, then distribution must also be recorded by weight. If acquisitions are recorded by physical count (e.g., by units), then distribution must also be recorded by physical count. [27 CFR 555.122-.125]

64. Must a licensee or permittee maintain a daily summary of magazine transactions?

Yes. After the initial inventory required by regulations has been taken, the inventory shall be entered in a record of daily transactions. Not later than the close of the next business day, each licensee and permittee shall record by manufacturer's name or brand name the total quantity received in and removed from each magazine during the day and the total remaining on hand at the end of the day. [27 CFR 555.127]

65. Where must a licensee or permittee keep the daily summary of magazine transactions?

The records must either be kept at each magazine or at one central location on the business premises, provided a separate record of daily transactions is maintained for each magazine. [27 CFR 555.127]

66. How can I obtain additional copies of ATF Forms?

Forms are available on-line at www.atf.gov/forms/5000.htm#explosives. Requests for forms should be mailed to the ATF Distribution Center, 7943 Angus Court, Springfield, Virginia 22153. You may also have forms mailed to you by submitting an on-line request at www.atf.gov/dcof/index.htm, or by telephoning your request to 703-455-7801. [27 CFR 555.21(b)]

67. Does a purchaser of black powder have to sign any forms at the time of purchase?

If 50 pounds or less of commercially manufactured black powder is being purchased, and the powder is intended to be used solely for sporting, recreational, or cultural purposes in antique firearms as defined in 18 U.S.C. 921(a)(16) or in antique devices exempt from the term "destructive device" in 18 U.S.C. 921(a)(4), no form is required. However, if the black powder is being purchased for any other purpose (regardless of quantity), the purchaser or other transferee must possess a Federal explosives license or permit. [18 U.S.C. 845(a)(5); 18 U.S.C. 926(c); 27 CFR 555.141(b), 555.26(a)]

68. Is there a requirement for licensees and permittees to make an annual inventory of explosive materials on hand?

Yes. An inventory is required to be taken at least once a year. [27 CFR 555.122-.125]

69. When must ATF Form 5400.4, "Limited Permittee Transaction Report (LPTR)", be executed?

Before distribution of explosive materials to a limited permittee, the licensee or permittee must obtain an executed ATF F 5400.4 from the limited permittee with an original unaltered and unexpired Intrastate Purchase of Explosives Coupon (IPEC) attached. Except when delivery of explosive materials is made by a common or contract carrier who is an agent of the limited permittee, the licensee or permittee must verify the identity of the holder of the limited permit by examining an identification document (as defined in 555.11) and noting on the ATF F 5400.4 the type of document presented. The licensee or permittee must complete the appropriate section on ATF F 5400.4 to indicate the type and quantity of explosive materials distributed, the license or permit number of the seller, and the date of the transaction. The licensee or permittee must sign and date the form. [27 CFR 555.126(b)]

70. Do the ATF Forms 5400.4 have to be maintained by the licensee or permittee making the sale?

Yes. One copy of ATF F 5400.4 must be retained by the seller as part of his permanent records in chronological order by date of disposition, or in alphabetical order by name of limited permittee. They must be maintained for a period of five years. [27 CFR 555.126]

71. May I keep computerized records?

Yes. See ATF Ruling 2007-1.

Storage

72. Who must comply with the storage requirements?

Except for those items and activities given exempt status under 18 U.S.C. 845 (also see 27 CFR 555.141), or exempted under 27 CFR 555.32, Special Explosives Devices, all persons who store explosive materials must store them in conformity with the provisions of Subpart K of the regulations, unless the person or the materials are exempt from regulation. [18 U.S.C. 842(j); 27 CFR 555.29, 555.141, 555.201(a)]

73. What are the classes of explosive materials for storage purposes?

There are 3 classes of explosive materials:

(a) High explosives (for example, dynamite, flash powders, and bulk salutes);

(b) Low explosives (for example, black powder, safety fuses, igniters, igniter cords, fuse lighters, and "display fireworks", except for bulk salutes); and

(c) Blasting agents (for example, ammonium nitrate-fuel oil and certain water gels). [27 CFR 555.202]

74. May a person store explosive materials in a residence or dwelling?

No. Storage of explosive materials in a residence or dwelling is prohibited. [27 CFR 555.208(b), 555.210(b), 555.211(b)]

75. What is the "Table of Distances"?

This table lists the minimum acceptable distances separating explosives magazines from inhabited buildings, passenger railroads, public highways, and other explosives magazines. The table is contained in 27 CFR 555.218.

76. When low and high explosives are stored together, how is the distance determined to meet the table of distance requirements?

The table of distances at 27 CFR 555.218 would be applied using the total weight of explosive materials in the magazine. [27 CFR 555.218]

77. Is it necessary to inspect my explosives magazines on a regular basis?

Yes. Any person storing explosives must inspect the magazines at least once every 7 days to determine whether there has been unauthorized entry or attempted entry into the magazines or unauthorized removal of the contents of the magazines. [27 CFR 555.204]

78. What are the requirements for making changes or additions to an approved storage facility?

Making changes in construction to an approved explosives magazine or adding a magazine requires that ATF be notified. However, mobile or portable type 5 magazines and magazines used for the temporary (under 24 hours) storage of explosive materials are exempt from this requirement. See 27 CFR 555.63 for details.

79. Is any type of black powder fuse exempt from storage requirements?

Yes, 3/32-inch and other external burning pyrotechnic hobby fuses are exempt from the requirements of Federal explosives laws and regulations. [18 U.S.C. 845(a)(4-5); 27 CFR 555.11: definition of "ammunition", 555.141(a)(4), 555.141(b)]

80. With the exception of 3/32-inch pyrotechnic safety fuse for use in small arms, must black powder fuses generally be stored in approved explosives magazines?

Yes. Generally igniter fuses, time fuses, blasting fuses, safety fuses, or other black powder fuses by whatever name known, must be stored in approved magazines.

81. Is smokeless powder designed for use in small arms ammunition subject to the explosives storage requirements?

Smokeless propellants designed for use in small arms ammunition are exempt from regulation under 18 U.S.C. Chapter 40 and the regulations in 27 CFR Part 555. However, it should be noted that persons engaged in the business of importing or manufacturing smokeless propellants must have a Federal explosives license. Additionally, smokeless propellant designed for use other than small arms ammunition is not exempt. Therefore, explosives products such as squibs, fireworks, theatrical special effects, or other articles that may be utilizing smokeless propellants are regulated and must be stored accordingly.

82. My office building, in which several company employees work during the day in connection with my explosives business, is located in the general area of my explosives magazine. Do the regulations and the Table of Distances apply to this building as an "inhabited building"?

No. A building such as an office building or repair shop which is part of the premises of an explosives business and is used by the business in connection with the manufacture, transportation, storage, or use of explosive materials is not considered to be an "inhabited building". [27 CFR 555.11: definition of "inhabited building", 555.218]

83. Am I required to notify my State or local authorities about my explosives storage magazines?

Yes. All persons who store explosive materials must notify the fire department having jurisdiction over the site where explosive materials are manufactured or stored. Notification must be made orally by the end of the day on which storage begins and in writing within 48 hours from the time storage began. The notification must include the type of explosive materials, magazine capacity, and the location of each storage site. [27 CFR 555.11: Definition of "authority having jurisdiction for fire safety", 27 CFR 555.201(f)]

84. What is the definition of a "case hardened shackle?"

Case hardening involves putting carbon (or a combination of carbon and nitrogen) into the surface of the steel to make it a high-carbon steel, which can be hardened by heat treatment. Only the outer skin gets hard in this manner. The center is still tough and malleable. This makes for a strong lock with a tough surface.

85. Can detonators be stored with detonating cord?

No. However, products which are manufactured with a detonator attached to the detonating cord as an integral part need not be disassembled and stored separately. [27 CFR 555.213]

86. Are there storage requirements for oxidizers, such as ammonium nitrate?

In general, no. However, when a magazine or bin containing ammonium nitrate is located within the sympathetic detonation distance of other explosives or blasting agents, it must be stored in accordance with the table of distances in 27 CFR 555.220.

87. Are State and local government agencies required to store their explosive materials in conformity with Federal storage regulations?

Yes. There is no exemption in the law or regulations for the storage of explosive materials by any State or political subdivision thereof. [18 U.S.C. 842(j), 845(a)(6); 27 CFR 555.141(a)(3), (a)(5)]

Fireworks

Fireworks are defined in the Federal explosives regulations as any composition or device designed to produce a visible or an audible effect by combustion, deflagration, or detonation. Fireworks are further divided into two broad classifications, consumer fireworks or display fireworks as defined at 27 CFR Part 555.11.

88. Are "consumer fireworks" subject to regulation under the Federal explosives laws?

No. The importation, distribution, and storage of fireworks defined as consumer fireworks are exempted from the provisions of the Federal explosives laws. However, because they contain pyrotechnic compositions classed by ATF as explosive materials, the manufacture of consumer fireworks requires a manufacturer's license. In addition, pyrotechnic compositions used in the manufacture of consumer fireworks must be stored in accordance with regulations in 27 CFR Subpart K. Consumer fireworks are defined as "any small firework device designed to produce visible effects by combustion and which must comply with the construction, chemical composition, and labeling regulations of the U.S. Consumer Product Safety Commission, as set forth in title 16, Code of Federal Regulations, parts 1500 and 1507. Some small devices designed to produce audible effects are included, such as whistling devices, ground devices containing 50 mg or less of explosive materials, and aerial devices containing 130 mg or less of explosive materials. Consumer fireworks are classified as fireworks UN0336 and UN0337 by the U.S. Department of Transportation at 49 CFR 172.101. This term does not include fused set pieces containing components which together exceed 50 mg of salute powder." [27 CFR 555.11: definition of "consumer fireworks"; definition of "licensed manufacturer", 555.141(a)(7)]

89. Are "display fireworks" considered to be explosive materials subject to regulation under Federal explosives laws and regulations?

Yes. Display fireworks include, but are not limited to, salutes containing more than 2 grains (130 mg) of explosive materials, aerial shells containing more than 40 grams of pyrotechnic compositions, and other display pieces which exceed the limits of explosive materials for classification as "consumer fireworks". These fireworks are classified as fireworks UN0333, UN0334, or UN0335 by regulations of the U.S. Department of Transportation at 49 CFR 172.101. Display fireworks also include fused set pieces containing components which together exceed 50 mg of salute powder. [27 CFR 555.11: definition of "display fireworks"]

90. How must display fireworks be stored?

Display fireworks, with the exception of bulk salutes, are considered low explosives and, at a minimum, must be stored in type 4 storage magazines. They may also be stored in type 1 or type 2 magazines. Bulk salutes, which are defined as either salute components prior to final assembly into aerial shells, (or) finished salute shells held separately prior to being packed with other types of display fireworks, are classified as high explosives. As such, bulk salutes may only be stored in type 1 or type 2 magazines specifically constructed for the storage of high explosives. [27 CFR 555.11, 555.202(b), 555.203(d), 555.207, 555.208, 555.210]

91. Are "Articles Pyrotechnic" subject to the requirements of the Federal explosives regulations?

The importation, distribution, and storage of fireworks defined as "Articles Pyrotechnic", are exempt from the Federal explosives laws and regulations. However, because they contain pyrotechnic compositions classed by ATF as explosive materials, the manufacture of items defined as "articles pyrotechnic" requires an ATF manufacturer's license. In addition, pyrotechnic compositions used in the manufacture of articles pyrotechnic must be stored in accordance with regulations in 27 CFR Subpart K. [27 CFR 555.11: definitions of "articles pyrotechnic" and "consumer fireworks", 555.141(a)(7)]

92. Must partially assembled display fireworks be removed from a drying building for overnight storage?

Yes. At the end of a day's manufacturing operations, all dry explosive powders and mixtures and partially assembled and finished display fireworks must be removed from fireworks process buildings and stored in a magazine meeting the storage requirements in 27 CFR Part 555, Subpart K. [27 CFR 555.205, 555.221]

93. What areas of a fireworks manufacturing plant are considered to be "fireworks process buildings?"

Fireworks process buildings include any buildings in which pyrotechnic compositions or explosives materials are mixed, pressed, finished, or assembled. Fireworks process buildings do not include plant warehouses, office buildings, or other buildings and areas in which no fireworks, pyrotechnic compositions, or explosive materials are processed or stored. [27 CFR 555.11: definition of "fireworks process building"]

94. Under what conditions may I temporarily store display fireworks (including low explosives for choreographed shows) on trucks?

See ATF Ruling 2007-2.

95. What types of fireworks require an ATF license or permit in order to be lawfully transported or received?

Any fireworks defined as "display fireworks" in 27 CFR 555.11 may be lawfully received or transported only by persons who hold a valid license or permit. No ATF license or permit is required to receive or transport "consumer fireworks" or "articles pyrotechnic". [18 U.S.C. 842(a)(3); 27 CFR 555.26, 555.141(a)(7)]

Plastic Explosives

96. What is a plastic explosive?

A plastic explosive is defined as "an explosive material in flexible or elastic sheet form formulated with one or more high explosives which in their pure form has a vapor pressure less than 10^{-4} Pa at a temperature of 250 C., is formulated with a binder material, and is as a mixture malleable or flexible at normal room temperature." [18 U.S.C. 841(q); 27 CFR 555.180(d)(4)]

97. What plastic explosives are required to contain detection agents?

All plastic explosives manufactured or imported on or after April 24, 1996, must contain a detection agent. Federal law enforcement agencies and the military may possess unmarked plastic explosives if they meet the requirements of the use-up period described in Question 103. [18 U.S.C. 841(q), 842(n); 27 CFR 555.180]

98. What are the permissible detection agents for marking plastic explosives?

These agents are listed in the law and regulations at 18 U.S.C. 841(p) and 27 CFR 555.180(d)(3).

99. Is it lawful to manufacture plastic explosives that do not contain a detection agent?

No. [18 U.S.C. 842(l); 27 CFR 555.180(a)]

100. Is it lawful to import into the United States plastic explosives that do not contain a detection agent?

No. The importation of plastic explosives into the United States requires that the importer file ATF Form 6 certifying that the imported plastic explosives contain the required detection agent, or is exempted from the marking requirements as provided in the regulations. [18 U.S.C. 842(m); 27 CFR 555.180(b), 555.182, 555.183]

101. Is it lawful to ship, transport, transfer, receive, or possess any plastic explosive that does not contain a detection agent?

No. However, a 15-year use-up period is provided for Federal law enforcement agencies and the military for unmarked plastic explosives imported into or manufactured in the U.S. prior to April 24, 1996. [18 U.S.C. 842(n); 27 CFR 555.180(c)]

102. If a person acquired plastic explosives not containing a detection agent before April 24, 1996, may he or she continue to lawfully possess the explosives?

No. With the exception of the use-up period provided by law for Federal law enforcement agencies or the military, the time period for lawful possession of unmarked plastic explosives terminated on April 24, 1999. [18 U.S.C. 842(n); 27 CFR 555.180(c)]

103. Are police departments exempt from the prohibition against possessing unmarked plastic explosives after April 24, 1999?

No. Police departments and other State or local law enforcement agencies could lawfully possess unmarked plastic explosives acquired on or before April 24, 1996, until April 24, 1999. Such agencies still possessing unmarked plastic explosives should destroy them or abandon them to ATF. Contact the nearest ATF field office for information. [18 U.S.C. 842(n); 27 CFR 555.180(c)(1)]

U.S. Military Explosives

104. Would an ATF license or permit be needed to demilitarize (demil) U.S. military explosives?

As long as the demil operator has a valid Department of Defense contract to perform such operations, the operations would be exempt from 27 CFR Part 555 and no license or permit would be required. However, if title to the explosive materials has passed from the military to the demil operator and the operator intends to resell the explosives on the commercial market, then such operations may be regulated by ATF (e.g., storage, sales, manufacturing) and an ATF license or permit may be needed. Contact the nearest ATF field office for further information. [18 U.S.C. 845(a)(3), (a)(6) and 27 CFR 555.141(a)(3), (a)(5)]

105. Would a civilian contractor who is manufacturing explosive materials pursuant to a government contract for or on behalf of the United States military be entitled to the exemptions from the explosives laws and regulations?

Yes, provided that all the explosive materials in question are manufactured under a government contract. Any explosive materials manufactured in anticipation of receiving a government contract would not qualify for this exemption.

If the contractor manufactures any explosive materials not pursuant to a U.S. military contract, the manufacture and the explosive materials are subject to all requirements of the law and regulations. [18 U.S.C. 845(a)(3), (a)(6); 27 CFR 555.26, 555.41, 555.141(b)]

106. Is an ATF licensee or permittee, whose licensed premises are located on a U.S. military installation, subject to the regulations in 27 CFR Part 555?

All activities conducted outside the scope of a U.S. Government contract are subject to the requirements of Part 555, even if the activities are conducted on property owned by the military. [18 U.S.C. 845(a)(3), (a)(6); 27 CFR 555.26, 555.29, 555.41, 555.141(a)(3), (a)(5)]

Index to Questions and Answers

ATF Explosives Rulings (Revised 1/07)

Table of Contents

1. 27 CFR 55.11: Meaning of Terms (Also § 55.206)

An office or repair shop used in connection with the manufacture, etc. of explosive materials is not an "inhabited building."

ATF Ruling 75-20

ATF has held that a building, such as an office or repair shop, which is a part of the premises of an explosives manufacturer and is used in connection with the manufacture, transportation, storage, or use of explosive materials, is not an "inhabited building." Section 55.11 of 27 CFR defines inhabited building as "any building regularly occupied in whole or in part as a habitation for human beings, or any church, schoolhouse, railroad station, store, or other structure where people are accustomed to assemble, except any building occupied in connection with the manufacture, transportation, storage, or use of explosive materials."

Regulations in 27 CFR §§ 55.206 and 55.218 set forth provisions concerning the location of storage facilities and the minimum distances such storage facilities may be located from, among other things, "inhabited buildings." These provisions are intended to provide protection to persons who inhabit buildings located near premises where explosives are manufactured, stored, etc. However, it is the intent of § 55.11 to exempt buildings used by the explosives industry in connection with the manufacture, transportation, storage, or use of explosive materials from the table of distance requirements on "inhabited buildings." [75 CB 64]

2. 27 CFR 55.207: Construction of Type 1 Magazines (Also § 55.210)

Certain explosives storage facilities meeting standards of construction prescribed by the Department of Defense Explosives Safety Board for such storage are approved by the Bureau.

ATF Ruling 75-21

ATF has held that explosives storage facilities with smooth-finished concrete floors that were constructed under contract for the use of the Department of Defense (DOD) and that are presently being leased to licensees and permittees for the storage of commercial explosives are considered to be in compliance with the requirements for nonsparking floors, as set forth in 27 CFR §§ 55.207(a)(4), 55.207(b), and 55.210, for the storage of all types of fully packaged explosives, pyrotechnics and propellants, with the exception of black powder. Any other such magazines which have smooth finished concrete floors and which meet or exceed DOD construction specifications will also be considered to be in compliance with the requirements of Part 555 with respect to nonsparking floors. It is the responsibility of the licensee or permittee to provide verification that such facilities were manufactured under DOD specifications or that the facilities meet or exceed such specification standards. If the Division Director determines that the concrete floors of type 1 or type 4 magazines do not meet the preceding requirements, he will require such floors to be covered with a nonsparking material, such as epoxy paint or mastic. [75 CB 67]

3. 27 CFR 55.41: Licenses and Permits-General

Certain companies that manufacture explosive materials for use in their own operations are required to obtain licenses as manufacturers of explosive materials.

ATF Ruling 75-31

ATF has held that companies, such as public utility companies engaged in line and facility construction, which manufacture explosives on a regular or continual basis are considered to be engaged in the business of manufacturing explosive materials and must be appropriately licensed as required by 18 U.S.C. 842. The term "manufacturer" is defined in 18 U.S.C. 841(h) as "any person engaged in the business of manufacturing explosive materials for purposes of sale or distribution or for his own use." Although the term "engaged in the business" is not susceptible to a rigid definition within 18 U.S.C. §§ 841-848, it is interpreted to imply an element of continuity or habitual practice; an element clearly present in the operations of companies described herein. Therefore, these companies are considered to be "engaged in the business" and must be licensed as explosives manufacturers. [75 CB 65]

4. 27 CFR 555.109: Identification of Explosive Materials

Methods of marking containers of explosive materials are prescribed.

ATF Ruling 75-35

Editors note: ATF Ruling 75-35 was rendered obsolete pursuant to ATF 5F, 70 Federal Register 30626 (May 27, 2005), and effective July 26, 2005.

5. 27 CFR 55.11: Meaning of Terms- State of Residence

"State of residence" of business entities who use explosive materials; distribution of explosive materials by licensees to out-of-State business entities other than licensees and permittees; and distribution to nonresident employees of such entities are discussed.

Editor's Note: Provisions of ATF Ruling 76-4 were modified, in part, by the Safe Explosives Act.

Effective May 24, 2003, it is unlawful for any person to receive explosive materials unless such person holds an ATF license of permit. It is also unlawful for any licensee or permittee to knowingly distribute explosive materials to any person who does not hold a license or permit. The only relevance remaining in the term "State of residence" is for distribution of explosive materials to, and receipt by, limited permit holders. Pursuant to 18 U.S.C. 842(a)(3) and (a)(4), limited permit holders may, on not more than 3 separate occasions, lawfully receive explosive materials from a licensee or permittee whose premises are located within the state of residence of the limited permit holder. ATF Rule 76-4 continues to apply in determining whether a limited permit holder has acquired a "State of residence" for purposes of receipt of explosives under 18 U.S.C. 842(a)(4)(B).

6. 27 CFR 55.126: Explosives Transaction Record

Under certain conditions, a single Form 5400.4 may be used to cover a series of deliveries.

Editor's Note: The provisions of ATF Rule 76-10 were rendered obsolete by ATF No. 1, 68 FR 13791, Mar. 20, 2003.

7. 27 CFR 55.207: Construction of Type 1 Storage Facilities (Also § 55.208)

Alternate construction standards for storage facilities for explosive materials are prescribed.

ATF Ruling 76-18

Section 842(j) of 18 U.S.C. states: "It shall be unlawful for any person to store any explosive material in a manner not in conformity with regulations promulgated by the Secretary. In promulgating such regulations, the Secretary shall take into consideration the class, type, and quantity of explosive materials to be stored, as well as the standards of safety and security recognized in the explosives industry." The regulations in 27 CFR §§ 55.207 and 55.208 prescribe types of storage facilities for explosive materials and provide (among other things) that such storage facilities shall be bullet resistant.

Section 55.201(b) provides that alternate storage facilities may be authorized for the storage of explosive materials when it is shown that such alternate facilities are or will be constructed in a manner substantially equivalent to the standards of construction contained in the applicable regulations.

The term "bullet-resistant" means resistant to penetration of a bullet of 150 grain M2 ball ammunition having a nominal muzzle velocity of 2700 feet per second fired from a .30 caliber rifle from a distance of 100 feet perpendicular to the wall or door. It has been determined that a wide range of construction criteria meet the bullet-resistant requirements of regulations for construction of storage facilities for explosive materials. In order to promote standards of safety and security in the storage of explosive materials while allowing the industry a wide latitude in the selection of construction materials, it is held that storage facilities (magazines) that are constructed according to the following minimum specifications are bullet-resistant and meet the requirements of the regulations as set forth in 27 CFR Part 55 (All steel and wood dimensions are actual thicknesses. To meet the concrete block and brick dimensions indicated, the manufacturers' represented thicknesses may be used).

(a) Exterior of ⅝ inch steel, lined with an interior of any type of nonsparking material.

(b) Exterior of ½ inch steel, lined with an interior of not less than ⅜ inch plywood.

(c) Exterior of ⅜ inch steel, lined with an interior of two inches of hardwood.

(d) Exterior of ⅜ inch steel, lined with an interior of three inches of softwood or 2¼ inches of plywood.

(e) Exterior of ¼ inch steel, lined with an interior of three inches of hardwood.

(f) Exterior of ¼ inch steel, lined with an interior of five inches of softwood or 5 ¼ inches of plywood.

(g) Exterior of ¼ inch steel, lined with an intermediate layer of two inches of hardwood and an interior lining of 1½ inches of plywood.

(h) Exterior of 3⁄16 inch steel, lined with an interior of four inches of hardwood.

(i) Exterior of 3⁄16 inch steel, lined with an interior of seven inches of softwood or 6¾ inches of plywood.

(j) Exterior of 3⁄16 inch steel, lined with an intermediate layer of three inches of hardwood and an interior lining of ¾ inch of plywood.

(k) Exterior of ⅛ inch steel, lined with an interior of five inches of hardwood.

(l) Exterior of ⅛ inch steel, lined with an interior of nine inches of softwood.

(m) Exterior of ⅛ inch steel, lined with an intermediate layer of four inches of hardwood and an interior lining of ¾ inch plywood.

(n) Exterior of any type of fire-resistant material which is structurally sound, lined with an intermediate layer of four inches of solid concrete block, OR four inches of solid brick OR four inches of solid concrete; AND, an interior lining of ½ inch plywood placed securely against the masonry lining.

(o) Standard eight inch concrete block with voids filled with well-tamped sand/cement mixture.

(p) Standard eight inch solid brick.

(q) Exterior of any type of fire-resistant material which is structurally sound, lined with an intermediate six inch space filled with well-tamped dry sand or well-tamped sand/cement mixture.

(r) Exterior of 1/8 inch steel, lined with a first intermediate layer of ¾ inch plywood, a second intermediate layer of 3⅝ inches of well-tamped dry sand or sand/cement mixture and an interior lining of ¾ inch plywood.

(s) Exterior of any type of fire-resistant material, lined with a first intermediate layer of ¾ inch plywood, a second intermediate layer of 3⅝ inches well-tamped dry sand or sand/cement mixture, a third intermediate layer of ¾ inch plywood, and a fourth intermediate layer of two inches of hardwood OR 14 gauge steel AND an interior lining of ¾ inch plywood.

(t) Eight inch thick solid concrete. [76 CB 106]

8. 27 CFR 55.213: Quantity and Storage Restrictions(Also § 55.208)

Alternate magazine construction standards for storage of electric blasting caps with other explosive materials are prescribed.

ATF Ruling 77-24

Section 842(j) of 18 U.S.C. states: "It shall be unlawful for any person to store any explosive material in a manner not in conformity with regulations promulgated by the Secretary. In promulgating such regulations, the Secretary shall take into consideration the class, type, and quantity of explosive materials to be stored, as well as the standards of safety and security recognized in the explosives industry." The regulations in 27 CFR § 55.213 restrict the storage of blasting caps with other explosive materials. Section 55.201(b) provides that alternate storage magazines may be authorized for the storage of explosive materials when it is shown that such alternate magazines are or will be constructed in a manner substantially equivalent to the standards of construction contained in the applicable regulations.

ATF recognizes that the transportation and storage of explosive materials in the same

vehicle along with electric blasting caps is often desired. The Institute of Makers of Explosives established a recommended standard for such transport in their Safety Library Publication No. 22, dated November 5, 1971 [revised January 1985]. This standard prescribes the minimum construction criteria for:

(a) A container securely attached—

(1) Above the cab of the vehicle (see Figure 1, Appendix A), and

(2) To the vehicle frame under the cargo space (see Figure 2, Appendix A), or

(b) A built-in compartment in the cargo space of the vehicle (see Appendix B).

In addition to motorized vehicles, consideration was also given for the use of similar criteria on portable wheeled trailers being used as magazines under § 55.208(a) of the regulations (see

Appendix E). In order to insure standards of safety and security in the storage of explosive materials while allowing the industry a proper latitude in the construction of magazines, it is held that vehicles used for transporting and for storing explosive materials that are constructed in conformity with the standards listed below, and in compliance with all other safety and security provisions contained in Part 55 (e.g., effectively immobilized when unattended) will meet the requirements of ATF regulations. Even though constructed on the same vehicle, each compartment will be considered as a separate magazine. The two magazines on the vehicle will, however, be considered as one magazine when applying the American Table of Distances [see Table at § 55.218].

Construction Standards For Storage of Electric Blasting Caps (Non Mass-Detonating)

a. The container or compartment must provide for total enclosure of the electric blasting caps.

b. The partition between the explosives storage compartment and the electric blasting cap compartment must be of laminate construction consisting of A/C grade or better exterior plywood, gypsum board [sheetrock] and low carbon steel plates. In order of arrangement, the laminate must conform to the following, with minimum thickness of each lamination as indicated:
½ inch plywood
½ inch gypsum board [sheetrock],
⅛ inch low carbon steel, and,
¼ inch plywood, with the ¼ inch plywood facing the explosives storage compartment. See Appendix C for details of laminate construction. The door to the electric blasting cap compartment must be of metal construction or solid wood covered with metal; the outside walls and top must be of the same construction as the rest of the vehicle or trailer. If high explosives or bullet sensitive explosive materials are stored in the vehicle, then the storage compartment of the vehicle must be constructed so as to be bullet-resistant.

c. As an alternative to the construction requirements shown in paragraph b, a container for use only as illustrated in Appendix A may be used when constructed as follows:

1. The top, lid or door, and the sides and bottom of each container must be of laminate construction consisting of A/C grade or better exterior plywood, solid hardwood, gypsum board [sheetrock], and sheet metal. In order of arrangement, the laminate must conform to the following, with minimum thickness of each lamination as indicated:
¼ inch plywood,
1 inch solid hardwood,
½ inch plywood,
½ inch gypsum board [sheetrock] (OR ¼ inch particle board), and 22 gauge sheet metal, constructed inside to outside in that order.
See Appendix D for details of laminate construction.

2. The hardwood must be fastened together with wood screws, the ½ inch plywood must be fastened to the hardwood with wood screws, the inner ¼ inch plywood must be fastened to the hardwood with adhesive, and the 22 gauge sheet metal must be attached to the exterior of the container with screws.

d. The laminate composite material must be securely bound together by waterproof adhesive or other equally effective means.

e. The steel plates at the joints of laminations must be secured by continuous fillet welds.

f. All interior surfaces of the container or compartment must be constructed so as to prevent contact of contents with any sparking metal.

g. There must be direct access to the container or into a compartment from outside the vehicle.

h. Each container or compartment must have a snug fitting continuous piano-type hinged lid or door equipped with a locking device (or devices).

i. Without permitting direct access to contents under normal conditions, the locking or hinging mechanisms must permit at least one edge of the lid or door to rise or move outward at least ½ inch when subjected to internal pressure.

j. The exterior of the container or compartment must be weather-resistant. [77 CB 191]

9. 18 U.S.C. 842(j): Storage of Explosives

27 CFR 55.208(b)(1), 55.210(b)(1), and 55.211(b)(1): Indoor Storage of Explosives in a Residence or Dwelling

ATF will approve variances to store explosives in a residence or dwelling only upon certain conditions including, but not limited to, receipt of a certification of compliance with State and local law, and documentation that local fire safety officials have received a copy of the certification.

ATF Ruling 2002-3

Section 842(j) of 18 U.S.C. states: "It shall be unlawful for any person to store any explosive material in a manner not in conformity with regulations promulgated by the Secretary."

The regulations in 27 CFR 55.208(b)(1), 55.210(b)(1), and 55.211(b)(1) specify that no indoor magazine is to be located in a residence or dwelling. Section 55.22 specifies that the Director may allow alternate methods or procedures in lieu of a method or procedure specifically prescribed in the regulations. Specifically, section 55.22(a)(3) provides that such "variances" are permissible only in certain circumstances, including where "[t]he alternate method or procedure will not be contrary to any provision of law and will not . . . hinder the effective administration of this part."

ATF has been advised that certain variances previously approved for storage of explosives in residences or dwellings are in violation of State or local zoning law. ATF believes it is important to ensure that approval of variances is in compliance with all State and local provisions.

To obtain a variance for indoor storage of explosives in a residence or dwelling, ATF has determined that a person must submit to ATF a certification signed under penalty of perjury along with the request for the variance. The certification must:

1. State that the proposed alternative storage method will comply with all applicable State and local law;

2. Provide the name, title, address, and phone number of the authority having jurisdiction for fire safety of the locality in which the explosive materials are being stored; and,

3. Demonstrate that the person has mailed or delivered the certification to the authority identified in (2).

When required by the Director, such persons must furnish other documentation as may be necessary to determine whether a variance should be approved.

Held, ATF will approve variances to store explosives in a residence or dwelling only upon certain conditions including, but not limited to, receipt of a certification of compliance with State and local law, and documentation that local fire safety officials have received a copy of the certification.

Date signed: August 23, 2002

10. 18 U.S.C. 842(j): Storage of Explosives 27 CFR 55.208(b)(1), 55.210(b)(1), and 55.211(b)(1): Indoor Storage of Explosives in Business Premises Directly Adjacent to a Residence or Dwelling

ATF Ruling 2002-4

ATF requires approval of variances for indoor storage of explosives in business premises directly adjacent to a residence or dwelling.

Section 842(j) of 18 U.S.C. states: "It shall be unlawful for any person to store any explosive material in a manner not in conformity with regulations promulgated by the Secretary."

The regulations in 27 CFR 55.208(b)(1), 55.210(b)(1), and 55.211(b)(1) specify that no indoor magazine is to be located in a residence or dwelling. Section 55.22 specifies that the Director may allow alternate methods or procedures in lieu of a method or procedure specifically prescribed in the regulations. Specifically, section 55.22(a)(3) provides that such "variances" are permissible only in certain circumstances, including where "[t]he alternate method or procedure will not be contrary to any provision of law and will not ... hinder the effective administration of this part."

ATF has been asked whether businesses that are directly adjacent to living quarters may lawfully store explosive materials in the business premises. The issue presented is whether the premises amount to a "residence or dwelling" within the meaning of the regulations cited above.

Even where the business premises are segregable from the living quarters by the existence of a door or a common wall, the business premises retain their character as a residence or dwelling. Accordingly, indoor storage of explosives in such premises is generally prohibited and can be allowed only pursuant to an approved variance.

Held, ATF requires approval of variances for indoor storage of explosives in business premises directly adjacent to a residence or dwelling. ATF may approve such variances upon receipt of all appropriate certification and other documentation as may be requested.

Date signed: August 23, 2002

11. 18 U.S.C. 842(f): Unlawful Acts, 27 CFR 555.105(b)(6)(iii): Distribution of Explosives to Limited Permittees

Distributors distributing explosive materials to holders of limited permits via common or contract carrier may verify receipt of the explosive materials by telephone, facsimile, e-mail, or other means within three business days of shipment in lieu of requiring the common or contract carrier to verify the identity of the person accepting delivery of the explosives. The distributor shall make a notation on ATF Form 5400.4 indicating whether the shipment was received and the date and time of the contact with the distributee.

ATF Ruling 2003-5

The Bureau of Alcohol, Tobacco, Firearms and Explosives (ATF) has received questions from the explosives industry regarding the requirement under 27 CFR 555.105(b)(6)(iii) that, effective May 24, 2003, a common or contract carrier hired by a Federal explosives licensee or permittee verify the identity of the distributee, note the type and number of the identification document, and provide this information to the distributor. The distributor is required to record this information on ATF Form 5400.4, Limited Permittee Transaction Report (LPTR). Industry members have informed ATF that this requirement places an undue burden on common and contract carriers. Drivers are concerned that verifying the identity of persons accepting delivery of explosive materials by examining an identification document and providing the identification information to the distributor will be overly time consuming. Drivers are also concerned that they could be held personally liable for delivering explosives to persons not authorized to receive them.

ATF imposed the verification requirements of section 555.105(b)(6)(iii) to ensure that when explosive materials are sold by a distributor to a holder of a limited permit and transported by a common or contract carrier hired by the distributor, the explosive materials are delivered only to a person authorized to receive them. ATF continues to believe it is important that sellers of explosive materials verify that such materials are delivered to persons authorized to receive them.

However, it was not ATF's intention to impose an undue burden on common or contract carriers.

Section 555.22, Title 27, CFR, provides that the Director may approve an alternate method or procedure in lieu of a method or procedure specifically prescribed in the regulations when he finds that:

(1) Good cause is shown for the use of the alternate method or procedure;

(2) The alternate method or procedure is within the purpose of, and consistent with the effect intended by, the specifically prescribed method or procedure and that the alternate method or procedure is substantially equivalent to that specifically prescribed method or procedure; and

(3) The alternate method or procedure will not be contrary to any provision of law and will not result in an increase in cost to the Government or hinder the effective administration of Part 555.

ATF finds that there is good cause to authorize a variance to the provisions of section 555.105(b)(6)(iii) due to the undue burden placed on common or contract carriers by the verification requirement. Accordingly, ATF authorizes the following alternate method or procedure to the identification verification requirements of section 555.105(b)(6)(iii):

The distributor shall, no later than three business days after shipment of the explosive materials, contact the distributee by telephone, facsimile, e-mail or any other means to ensure that the shipment has been received. The distributor shall make a notation on ATF Form 5400.4 indicating whether the shipment was received and the date and time of the contact with the distributee.

ATF finds that the above alternate method is consistent with the verification provisions of section 555.105(b)(6)(iii), because it will ensure that delivery has taken place and document the information in the distributor's records.

The alternate method is not contrary to any provision of law, will not increase the costs to ATF, and will not hinder the effective administration of the regulations in 27 CFR Part 555.

Held, pursuant to 27 CFR 555.22, ATF authorizes a variance from the requirements of 27 CFR 555.105(b)(6)(iii) for Federal explosives licensees and permittees making distributions of explosive materials to holders of limited permits via common or contract carrier. As an alternate method or procedure, distributors distributing explosive materials to holders of limited permits via common or contract carrier may verify receipt of the explosive materials by telephone, facsimile, e-mail or other means within three business days of shipment in lieu of requiring the common or contract carrier to verify the identity of the person accepting delivery of the explosives. The distributor shall make a notation on ATF Form 5400.4 indicating whether the shipment was received and the date and time of the contact with the distributee.

Date signed: May 23, 2003.

12. 27 CFR 555.11: Meaning of Terms

ATF provides guidance on three different private roads and whether they are "highways" as defined in 27 CFR 555.11.

ATF Ruling 2005-2

The Federal explosives laws, 18 U.S.C. Chapter 40, require all persons to store explosive materials in a manner in conformity with regulations issued by the Attorney General. 18 U.S.C. 842(j). The Attorney General has delegated his authority to administer and enforce the Federal explosives laws to the Director, ATF. 28 CFR 0.130. Regulations in 27 CFR Part 555 implement the provisions of the Federal explosives laws. Storage regulations in 27 CFR Part 555, Subpart K, provide that outdoor magazines in which high explosives are stored must be located no closer to inhabited buildings, passenger railways, public highways, or other magazines in which high explosives are stored than the minimum distances specified in the table of distances for storage of explosive materials in section 555.218 of the regulations. 27 CFR 555.206.

Section 555.11 of the regulations defines the term "highway" as "any public street, public alley, or public road, including a privately financed, constructed, or maintained road that is regularly and openly traveled by the general public."

In Scenario A, a private road owned by a corporation is used by the general public as an access road to a parking lot owned by the corporation. The road is near an explosives magazine. The road does not have a gate, sign, or other means of restricting access to the road. The road is also used by the general public on a daily basis to gain access to other public streets.

In Scenario B, a company that manufactures display fireworks, a logging company, and an individual who owns buildings utilized to store his collection of automobiles all occupy property to which the only access is a privately owned road. A separate party that leases to these three entities owns the property. The road is located on private property, and a locked gate at the entrance to the road prevents access by the general public. The display fireworks company, the logging company, and the individual storing automobiles all have keys to unlock the gate and travel on the road when needed. The gate is locked at all times, and there is no evidence that the road is open to anyone other than the two businesses and one individual who require access to their property.

In Scenario C, an explosives company maintains explosives magazines in a quarry area that has a roadway traversing through the quarry. The quarry owns the property, and the road is maintained by the quarry. The road has a gate and there are signs advising no trespassing. However, when ATF officials visited the location on several occasions, the gate was left open and members of the public regularly utilized the roadway as a shortcut between two major highways. There were no indications the owner of the property took any steps to prevent members of the public from utilizing the roadway.

Applying the regulatory definition of "highway" to the three scenarios, the road in Scenario A is clearly a highway that is subject to the tables of distance in Part 555. Although it is privately owned, it is regularly and openly traveled by members of the general public without restriction.

The roadway in Scenario B is not a "highway" as defined. Access is restricted at all times and there is no evidence the general public regularly travels on the roadway. Access to the road is provided to only a limited number of persons who have a legal right to travel the road. Accordingly, this road is not regularly and openly traveled by members of the general public.

ATF concludes that the roadway described in Scenario C is a "highway" as defined in 27 CFR 555.11. Although access to the roadway is restricted by a gate and "No trespassing" signs are posted, the gate is not closed at all times. Furthermore, ATF observation indicates that the roadway is regularly and openly traveled by members of the general public. Based on these facts, the roadway is a highway which is subject to the tables of distance in 27 CFR Part 555.

Held, a private road with no gate, signs, or other means of restricting access that is used by the general public as an access road to a parking lot and as access to other public streets is a "highway" as defined in 27 CFR 555.11.

Held further, a private road with a locked gate at the entrance that is locked at all times and used by a limited number of persons leasing or owning property accessed by the road is not a "highway" as defined in 27 CFR 555.11.

Held further, a private roadway traversing a quarry with a gate restricting access and a "no trespassing" sign is a "highway," as defined in 27 CFR 555.11, because the gate is not locked at all times and the general public regularly utilizes the roadway as a shortcut between two public highways.

Date approved: September 8, 2005

13. 27 CFR 555.11: Meaning of Terms

ATF provides guidance on two situations involving structures and whether they are "inhabited buildings" as defined in 27 CFR Part 555.

ATF Ruling 2005-3

The Federal explosives laws, 18 U.S.C. Chapter 40, require all persons to store explosive materials in a manner in conformity with regulations issued by the Attorney General. 18 U.S.C. 842(j). The Attorney General has delegated his authority to administer and enforce the Federal explosives laws to the Director, ATF. 28 CFR 0.130. Regulations in 27 CFR Part 555 implement the provisions of the Federal explosives laws. Storage regulations in 27 CFR Part 555, Subpart K, provide that outdoor magazines in which high explosives are stored must be located no closer to inhabited buildings, passenger railways, public highways, or other magazines in which high explosives are stored than the minimum distances specified in the table of distances for storage of explosive materials in section 555.218 of the regulations. 27 CFR 555.206.

The regulation at 27 CFR 555.11 defines the term "inhabited building" as "any building regularly occupied in whole or in part as a habitation for human beings, or any church, schoolhouse, railroad station, store, or other structure where people are accustomed to assemble, except any building occupied in connection with the manufacture, transportation, storage, or use of explosive materials."

In Scenario A, an explosives licensee leases explosives magazines to an individual who uses the magazines for storage of goods other than explosive materials. The magazines are located adjacent to magazines used by the explosives licensee for the storage of explosive materials. The magazines used by the lessee are not separated by the minimum distances required for the separation of magazines from "inhabited buildings" as required by the regulations in 27 CFR Part 555.

The magazines are visited regularly by the individual who stores property in the magazines, but no additional persons accompany the individual when he is present at the magazine. However, the individual hires contractors to repair equipment stored in the facilities from time to time, and 1-3 employees of the contractor may occasionally be present for short periods of time at the storage site. However, such visits occur no more than 3-5 times per year. In addition, an employee from the water company visits the storage site once a month to read the water meter, an employee from the power company reads the power meter once a month, and other vendors may be present at the site for short periods of time for other legitimate purposes.

In Scenario B, a licensed manufacturer of explosives X leases a unit in an industrial park that shares a common wall with a unit leased by licensed manufacturer Y. Both licensees store explosives in magazines located inside and outside the units. The magazines of Manufacturer X and the building used by Manufacturer Y are not separated by the minimum distance prescribed in 27 CFR Part 555 for the separation of magazines and inhabited buildings. Likewise, the magazines of Manufacturer Y and the building used by Manufacturer X are not separated by the minimum distance prescribed in 27 CFR Part 555 for the separation of magazines and inhabited buildings. In both structures, employees and contractors are regularly present during work hours for purposes of carrying on the manufacturing and distributing businesses of the two licensees. This includes personnel who work in the manufacturing plant, those who work on the loading dock to load and ship explosives products to customers, and those who work in the office taking orders, sending out invoices, and handling other clerical work for the businesses.

Applying the law and regulations to the facts of Scenario A, ATF concludes that the leased structures used by the individual to store items other than explosives are not being used as a habitation for human beings and are not buildings occupied in connection with the manufacture, transportation, storage, or use of explosive materials. Accordingly, the sole issue remaining is whether the structure is one where people are accustomed to assemble.

Noteworthy, the regulation uses the term "people," which is the plural version of "person." Thus, ATF believes that more than one person must "assemble" at the structure for it to be an "inhabited building." In addition, the word "assemble" is defined, in part, as "To bring or gather together in a group or whole." The word "assembly" is defined, in part, as "A group of persons gathered for a common purpose." The American Heritage Dictionary, Second College Edition, Houghton Mifflin Co., 1982. It is clear that the presence of one person at a structure or location cannot be an assembly of any sort. Accordingly, in situations where one person is present at a particular structure, whether on a regular or infrequent basis, such a structure is not an "inhabited building" as defined in 27 CFR 555.11.

Likewise, occasional visits to the storage facility by mail delivery persons or employees of public utility companies for brief periods of time would not be an "assembly" that would make the facility an inhabited building. However, if 2 or more repair persons are present at the facility to make repairs to equipment stored there, such persons would be there for a common purpose, and would have "assembled" at the structure. However, the structure would be an "inhabited building" only if it is a structure where people are accustomed to assemble. The word "accustom" is defined as "To familiarize, as by constant practice, use, or habit: accustomed himself to working long hours." The word "accustomed" is defined as "Usual, characteristic, or normal: worked with her accustomed thoroughness." The American Heritage Dictionary, Second College Edition, Houghton Mifflin Co., 1982. These definitions indicate that a structure will be one where people are accustomed to assemble only if there is some degree of continuity, regularity, or frequency to such assembly.

Infrequent, occasional visits to the storage site by 2 or more repair persons would not make the storage facility an "inhabited building," because such intermittent visits would not be "customary." Only where 2 or more persons are present at the site for a common purpose and on a regular basis would the building fit within the definition of "inhabited building."

To address Scenario B, it is apparent that the building leased by Manufacturer X is exempt from the definition of "inhabited building" as to the magazines of Manufacturer X, and the building leased by Manufacturer Y is likewise exempt as to the magazines of Manufacturer Y. This is because both buildings are occupied in connection with the manufacture, transportation, and storage of explosive materials. Clearly, the employees of both licensees are aware that explosive materials are present on the premises and they assume the risk of any such operation. A more difficult question is presented by the buildings of Manufacturer X and the magazines of Manufacturer Y and vice versa. ATF cannot assume that all employees are cognizant of the activities of their neighbors in the industrial park. Thus, it cannot be assumed that the employees are knowingly assuming the risk of explosive materials stored in magazines owned by the other licensee.

Given the plain language of the regulation, however, ATF does not believe it is appropriate to deny the coverage of the regulatory exemption to adjoining licensees on the basis of an assumption of the risk analysis. The current regulatory definition excludes from the definition of "inhabited building" any building occupied in connection with the manufacture, transportation, storage, or use of explosive materials, regardless of the knowledge of the building's occupants. Accordingly, ATF concludes that the industrial units occupied by Manufacturer X and Manufacturer Y are both exempted from the definition of "inhabited building" as to the magazines of each other as well as to their own magazines.

Held, a structure used to store items other than explosive materials that is visited on a regular basis by one individual is not an "inhabited building" as defined in 27 CFR 555.11, because it is not a structure where people are accustomed to assemble. Where 2 or more persons are present at the structure to repair equipment stored therein and such visits to the site are occasional and infrequent, the structure is not an "inhabited building" because the visits are not "customary." However, where 2 or more persons make regular visits to the structure for a common purpose, the structure is an "inhabited building," and explosives magazines may not be stored closer to the structure than the minimum distances specified in the regulations in 27 CFR Part 555.

Held further, buildings occupied by licensed explosives manufacturers in connection with the manufacture, transportation, storage, or use of explosive materials are not included within the definition of "inhabited building" as to magazines located on their own premises. In addition, buildings occupied by licensed explosives manufacturers in connection with the manufacture, transportation, storage, or use of explosives are not included within the definition of "inhabited building" as to magazines located on property owned by another licensee.

Date approved: November 25, 2005

14. 18 U.S.C. 842(f): Records Required for Explosives Licensees and Permittees

27 CFR 555, Subpart G: Records and Reports

27 CFR 555.22: Alternate Methods or Procedures; Emergency Variations from Requirements

Under specified conditions, approval is granted to utilize computerized records as required records under 27 CFR 555, Subpart G.

ATF Ruling 2007-1

Section 842(f), Title 18, United States Code, makes it unlawful for any licensee or permittee to willfully manufacture, import, purchase, distribute, or receive explosive materials without making such records as the Attorney General may by regulation require, including, but not limited to, a statement of intended use, the name, date, place of birth, social security number or taxpayer identification number, and place of residence of any natural person to whom explosive materials are distributed.

Regulations implementing section 842(f) are in 27 CFR Part 555, Subpart G. The regulations in this subpart specify the records required to be created and maintained by licensed importers (section 555.122), licensed manufacturers (section 555.123), licensed dealers (section 555.124), and permittees (section 555.125). The regulation in section 555.121 provides that licensees and permittees must keep records pertaining to explosive materials in permanent form (i.e., commercial invoices, record books) and in the manner required in Subpart G. In addition, sections 555.122-555.125 specifically allow licensees and permittees to use an alternate record to record the distribution of explosive materials when it is shown that the alternate records would accurately and readily disclose the information required by the regulations. These regulations require licensees and permittees who propose to use alternate records to submit a letter application to ATF describing the proposed alternate records and the need for them. Alternate records are not to be employed until approval from ATF is received.

Regulations at 27 CFR 555.22 allow for the approval and use of an alternate method or procedure in lieu of a method or procedure specifically prescribed in Part 555. ATF may approve an alternate method or procedure when it is found that—

(1) Good cause is shown for the use of the alternate method or procedure;

(2) The alternate method or procedure is within the purpose of, and consistent with the effect intended by, the specifically prescribed method or procedure and that the alternate method or procedure is substantially equivalent to that specifically prescribed method or procedure; and

(3) The alternate method or procedure will not be contrary to any provision of law and will not result in an increase in cost to the Government or hinder the effective administration of 27 CFR Part 555.

With advances in technology and the dramatic decrease in the cost of computers, many businesses rely upon computers to maintain records of their inventory, sales, customer lists, and other business information. Even the smallest home-based business utilizes computers to record and maintain business information. Creating and maintaining records in a computer database, rather than paper form, makes it easier to ensure accuracy of records and makes it less likely that records will be lost or misplaced. In addition, maintaining records via computer generally saves time and money in bookkeeping and auditing expenses. This utilization of computers has allowed companies to automate inventories, utilizing technology such as bar codes or RFID (radio frequency identification) chips. This facilitates better accountability of product overall, reducing the potential of everyday accounting errors. Over the years ATF has seen a significant increase in the number of requests from explosives licensees and permittees for authorization to utilize computerized records rather than paper records of acquisition and distribution and other required records, such as magazine transaction records. ATF routinely approves requests to utilize computerized records, with certain conditions, finding that the use of such records is substantially equivalent to methods of record keeping set forth in the regulations in 27 CFR Part 555, Subpart G.

Several explosives industry members have asked whether computerized records may be maintained without obtaining written approval from ATF if they contain all the required information specified in the regulations and are maintained in a permanent form. Additionally, industry members have questioned whether computer records in combination with paper records may be maintained if they are permanent and contain all the information required by the regulations.

ATF has determined that records of acquisition and disposition, magazine summary records, and the other records required by 27 CFR Part 555, Subpart G, satisfy the standard of permanency and are substantially equivalent to paper records if they meet the following criteria:

1. All data entered into the computer system must be recorded into the database and cannot be capable of being edited or modified at a later date. The software system must retain any correction of errors as an entirely new entry, without deleting or modifying the original entry. The system may allow for entries in a notes column to explain any correction.

2. The system must have a reliable daily memory backup capability to protect the data from accidental deletion or other system failure.

It is also acceptable for licensees/permittees to maintain required records using a combination of a computer program, commercial invoices, and other documents, provided that all of the required information is maintained in the records in permanent form. Any use of a computer for any portion of the required records must comply with the standards outlined above. However, each particular transaction must be self-contained with all the required information in the same recordkeeping medium. As one example, dispositions of explosives by a dealer cannot be separated by keeping the dates of disposition and the manufacturer's name or brand name in the computer, and all the other required information for that disposition on separate written documents.

ATF finds that good cause exists for authorizing the use of a computer to create and maintain the records required by 27 CFR Part 555, Subpart G, as the use of computers is accepted throughout the business community as a reliable, cost-efficient means of maintaining business records. ATF also finds that the use of a computer to maintain required records, contingent upon the requirements outlined above, is consistent with the effect intended by the requirements of Subpart G, as it will result in a permanent, reliable record that will accurately indicate acquisitions and dispositions of explosive materials.

Finally, ATF finds that the use of computer records properly containing all the required information should not hinder the effective administration of the Federal explosives laws or regulations – use of such records generally makes it easier for ATF to conduct inventories of product on hand and to audit required records. Accordingly, ATF concludes that the requirements for approval of an alternate method or procedure in accordance with 27 CFR Part 555, sections 555.22 and 555.122-555.125, are met.

Held, persons holding licenses and permits issued under 18 U.S.C., Chapter 40, may use computers to create and maintain all or any portion of the records required by 18 U.S.C. 842(f) and 27 CFR Part 555, Subpart G, if the following conditions are satisfied:

1. All data entered into the computer system must be recorded into the database and cannot be capable of being edited or modified at a later date. The software system must retain any correction of errors as an entirely new entry, without deleting or modifying the original entry. The system may allow for entries in a notes column to explain any correction.

2. The system must have a reliable daily memory backup capability to protect the data from accidental deletion or other system failure.

Held further, licensees and permittees who wish to use computers to create and maintain all or a portion of their required records in accordance with the requirements set forth in this ruling are not required to obtain advance approval in accordance with 27 CFR 555.22 or 555.121-125.

Held further, licensees and permittees utilizing a combination of a computer program, commercial invoices, and other paper documents as required records must ensure that the required information for a particular transaction is fully contained in the same recordkeeping medium.

Held further, this ATF approved alternate method or procedure for computerized records shall not be withdrawn unless the holder of said variance is so advised by ATF in writing or no longer holds a Federal explosives license or permit.

Date approved: January 18, 2007

15. 18 U.S.C. 842(j): Storage of Explosives

27 CFR 555.210: Construction of Type 4 Magazines

27 CFR 555.215: Housekeeping

27 CFR 555.201: Notification of Local Fire Officials

27 CFR 555.63: Explosives Magazine Changes

27 CFR 555.22: Alternative Methods or Procedures; Emergency Variations from Requirements

Under specified conditions, display fireworks may be temporarily stored in locked and attended motor vehicles at the explosives magazine site(s) and at fireworks display site(s) without meeting the locking requirements of 27 CFR 555.210 provided certain additional security measures are in place. Additionally, allowance per 27 CFR 555.215 is made for the fuel tanks containing volatile materials that may be on the temporary storage vehicles. Finally, slight variation is provided for notification requirements to ATF and local fire officials.

ATF Rule. 2007-2

The Federal explosives laws, 18 U.S.C. Chapter 40, require all persons to store explosive materials in a manner in conformity with regulations issued by the Attorney General. 18 U.S.C. 842(j). The Attorney General has delegated the authority to administer and enforce the Federal explosives laws to the Director, ATF. 28 CFR 0.130. Regulations in 27 CFR Part 555, implement the provisions of the Federal explosives laws.

The regulation at 27 CFR 555.210(a) states, in part, "[o]utdoor magazines are to be fire-resistant, weather-resistant, and theft-resistant." Partly to satisfy the theft-resistant requirement, this section requires that each door be equipped with two mortise locks; two padlocks fastened in separate hasps and staples; a combination of a mortise lock and a padlock; a mortise lock that requires two keys to open; or a three-point lock. Padlocks must have at least five tumblers and a case-hardened shackle of at least

⅜-inch diameter. Padlocks must be protected with not less than ¼-inch steel hoods constructed so as to prevent sawing or lever action on the locks, hasps, and staples.

The regulation at 27 CFR 555.215 states, in part, "[v]olatile materials are to be kept a distance of not less than 50 feet from outdoor magazines."

The regulation at 27 CFR 555.201 requires, in part, that any person storing explosive materials notify local fire authorities orally before the end of the day on which storage of the explosive materials began and in writing within 48 hours from the time such storage began.

The regulation at 27 CFR 555.63 requires that any licensee or permittee who acquires (adds) a storage magazine must notify ATF at least five business days in advance of using any added explosives storage magazine.

Regulations at 27 CFR 555.22 allow for the approval and use of an alternate method or procedure in lieu of a method or procedure specifically prescribed in Part 555. ATF may approve an alternate method or procedure when it is found that—

(1) Good cause is shown for the use of the alternate method or procedure;

(2) The alternate method or procedure is within the purpose of, and consistent with the effect intended by, the specifically prescribed method or procedure and that the alternate method or procedure is substantially equivalent to that specifically prescribed method or procedure; and

(3) The alternate method or procedure will not be contrary to any provision of law and will not result in an increase in cost to the Government or hinder the effective administration of 27 CFR Part 555.

ATF has approved a significant number of variances for temporary storage for a specified amount of time before a display fireworks event, as well as during and after the event until the remaining explosive materials can be placed back into the appropriate storage magazine.

Preparation of display fireworks shows and the transportation of explosive materials to numerous show sites often take place over a period of several days. Preparing and temporarily storing the fireworks for these shows ordinarily take place on delivery trucks and trailers in one storage location where the proprietor already maintains storage of explosives materials with a high degree of security and safety by complying with the provisions of 27 CFR Part 555.

Many display fireworks shows also take several days to prepare at the show site. During preparation and after the show is completed, explosive materials frequently must be temporarily stored. This is often either extra product that was brought to the show or misfires that have been maintained and must be returned to permanent storage.

Allowing flexibility through alternate methods or procedures for specific regulations increases both safety and security at these show sites. These procedures are needed to increase public safety, as well as facilitate smooth operations for the display fireworks industry. The highest risk of incidents involving the accidental ignition of display fireworks is during handling, with the next highest risk being transportation. Providing no flexibility to allow storage in the delivery vehicles would require the industry to dangerously load and unload from storage magazines to vehicles and back into a storage magazine. Additionally, ATF believes that providing this guidance allows for preplanning by the proprietor and consistency of regulatory application nationwide.

One of the major dangers around explosives is fire. Therefore, the regulations require that volatile materials be maintained a distance of not less than 50 feet from outdoor explosives storage magazines. ATF believes that requiring attended storage for display fireworks temporarily stored in vehicles will ensure public safety, in lieu of the 50 foot separation requirement. The attendee should be able to alert the proper authorities if needed to ensure that a fire does not compromise this storage, or may relocate these temporary storage magazines to a safe location away from an identified fire.

Held, ATF will approve alternate methods or procedures for the temporary storage of display fireworks in locked and attended vehicles at explosives magazine site(s), as well as at the fireworks display site(s), under the following conditions:

1. The doors to each storage compartment containing explosive materials must be locked with at least one steel padlock having at least five tumblers and a casehardened shackle of at least ⅜-inch diameter. The padlock does not need to be protected by a steel hood. However, each temporary storage magazine must be attended at all times for security purposes. The vehicle is considered "attended" when an authorized individual is within 100 feet of all temporary storage and has an unobstructed view of the vehicle(s) containing the explosive materials. The individual must remain awake and observant of activities around the vehicle(s).

2. The person who temporarily stores the explosive materials must notify in writing the authority having jurisdiction for fire safety in the locality in which the explosive materials are stored no less than 3 Federal office business days prior to utilizing the additional temporary storage magazine(s).

3. The person who temporarily stores the explosive materials must notify ATF in writing of the location of this storage no less than 3 Federal office business days prior to utilizing the additional temporary storage magazine(s).

All other provisions of 27 CFR Part 555 must be complied with as prescribed.

Held further, this ATF-approved alternate method or procedure for the temporary storage of display fireworks in locked and attended vehicles shall not expire unless the holder of said variance is so advised by ATF or no longer holds a Federal explosives license or permit.

Date approved: January 18, 2007

16. 18 U.S.C. 842(j): Storage of Explosives

27 CFR 555.211(a): Immobilization of Outdoor Type 5 Mobile Storage Magazines

27 CFR 555.215: Housekeeping

27 CFR 555.22: Alternative Methods or Procedures; Emergency Variations from Requirements

Under specified conditions, blasting agents may be stored in mobile type 5 magazines (bulk delivery trucks) without meeting the prescribed immobilization requirements of 27 CFR 555.211.

ATF Rule 2007-3

The Federal explosives laws, 18 U.S.C. Chapter 40, require all persons to store explosive materials in a manner in conformity with regulations issued by the Attorney General. 18 U.S.C. 842(j). The Attorney General has delegated his authority to administer and enforce the Federal explosives laws to the Director, ATF. 28 CFR 0.130. Regulations in 27 CFR

Part 555 implement the provisions of the Federal explosives laws.

The regulation at 27 CFR 555.211(a)(1) states, in part, "[o]utdoor magazines are to be weather-resistant and theft-resistant." This section further states, "[w]hen unattended, vehicular magazines must have wheels removed or otherwise be effectively immobilized by kingpin locking devices or other methods approved by the Director."

The regulation at 27 CFR 555.215 states, in part, "[v]olatile materials are to be kept a distance of not less than 50 feet from outdoor magazines."

The regulation at 27 CFR 555.22 allows for the approval and use of an alternate method or procedure in lieu of a method or procedure specifically prescribed in Part 555. ATF may approve an alternate method or procedure when it is found that—

(1) Good cause is shown for the use of the alternate method or procedure;

(2) The alternate method or procedure is within the purpose of, and consistent with the effect intended by, the specifically prescribed method or procedure and that the alternate method or procedure is substantially equivalent to that specifically prescribed method or procedure; and

(3) The alternate method or procedure will not be contrary to any provision of law and will not result in an increase in cost to the Government or hinder the effective administration of 27 CFR Part 555.

ATF has approved a significant number of variances for an alternate means of immobilizing preloaded bulk delivery vehicles when some additional security measures were put in place.

Bulk delivery vehicles are routinely utilized for on-site delivery of blasting services. Often these trucks contain a blasting agent as defined under 27 CFR 555.11. Utilization of these bulk products delivered on-site and used immediately has increased safety and security by reducing the number of remotely located storage trailers containing packaged blasting agent products. These delivery vehicles generally leave the explosives storage locations at unusual times of day or night, and for safety reasons they are loaded the day before for next day's delivery. These trucks are incapable of being disabled by a kingpin locking device, and the requirement to remove the wheels for immobilization is obviously not feasible. Additionally, there are times when explosive material remains on the vehicle when it returns from use. Most of these products degrade during handling, therefore, removing them from the bulk vehicle would not be an option.

One of the major dangers surrounding explosives is fire. Therefore, the regulations at 27

CFR 555.215 require that volatile materials be maintained a distance of not less than 50 feet from outdoor explosive storage magazines. Because these preloaded storage vehicles contain a fuel tank filled with volatile materials, they must remain in an area protected from fire such as gravel, paved, or closely mowed designated parking area.

ATF believes that the following alternate method of operation is substantially equivalent to the prescribed methods. Increased safety for employees and the public provides good cause for this alternate method. It is not contrary to law and will not result in any increased cost to the Government. Overall, ATF believes allowing for this flexibility assists with the effective administration of 27 CFR Part 555.

Held, ATF will approve alternate methods or procedures for the preloading and temporary storage of bulk blasting agents in delivery vehicles at explosive magazine site(s), when the security and immobilization meets the following criteria:

1. All doors on the vehicle are locked, the ignition key is removed, and the key is secured away from the truck.

2. When the site is not in operation, outer perimeter security is established. This may be by a variety of means such as a locked gate, security guards, fence, natural features, or a combination of these.

3. Each potential access point to explosive materials on a storage vehicle will be secured with a minimum of one padlock that has at least five tumblers and casehardened shackle of at least ⅜-inch diameter.

4. Each vehicle shall be immobilized through the use of a steering wheel locking device, lockable battery disconnect switch, or both.

5. All vehicles preloaded with blasting agents shall be parked in a company designated area not susceptible to fire propagation such as bare dirt, gravel, rock, paving, or closely mowed parking lot.

All other provisions of 27 CFR Part 555 must be complied with as prescribed.

Held further, licensees and permittees who wish to use the alternate method or procedure set forth in this ruling are not required to obtain advance approval in accordance with 27 CFR 555.22.

Held further, this ATF-approved alternate method or procedure for the temporary storage of bulk blasting agent products in locked and properly secured vehicles shall not be withdrawn unless the holder of said variance is so advised by ATF in writing or no longer holds a Federal explosives license or permit.

Date approved: January 26, 2007

APPENDIX A

PERMANENTLY MOUNTED CONTAINERS

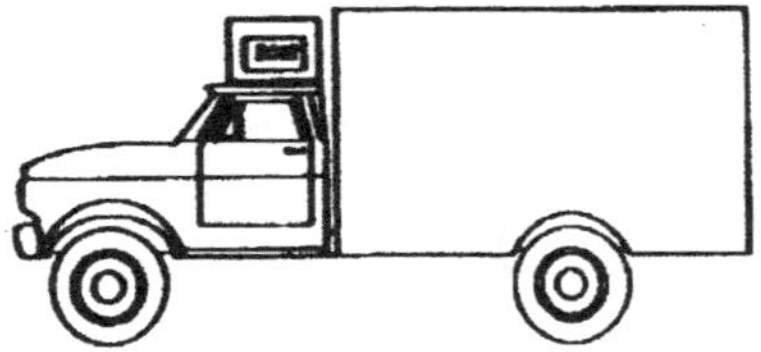

Figure 1

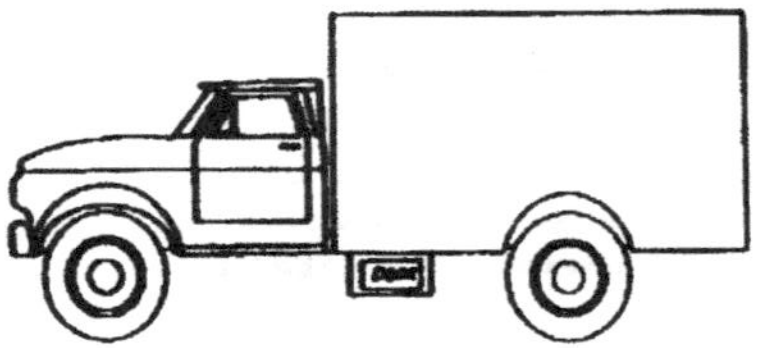

Figure 2

NOTE: The configurations shown in Figures 1 and 2 are equally applicable to multi-axle and "cab-over" vehicles.

[Diagrams: Courtesy of IME]

APPENDIX B

COMPARTMENTS

Figure 1

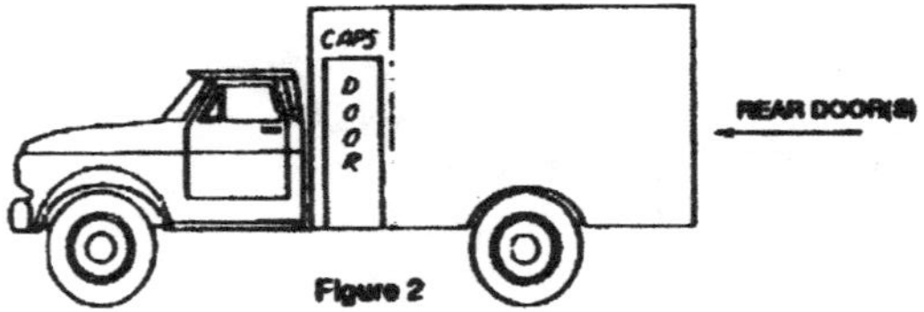

Figure 2

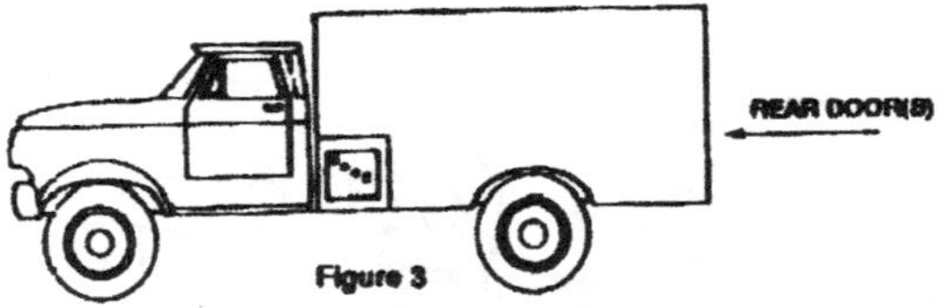

Figure 3

NOTE: The configurations shown in Figures 1 and 2 are equally applicable to multi-axle and "cab-over" vehicles.

[Diagrams: Courtesy of IME]

ATF Rul. 77–24, Appendix A and B

APPENDIX C

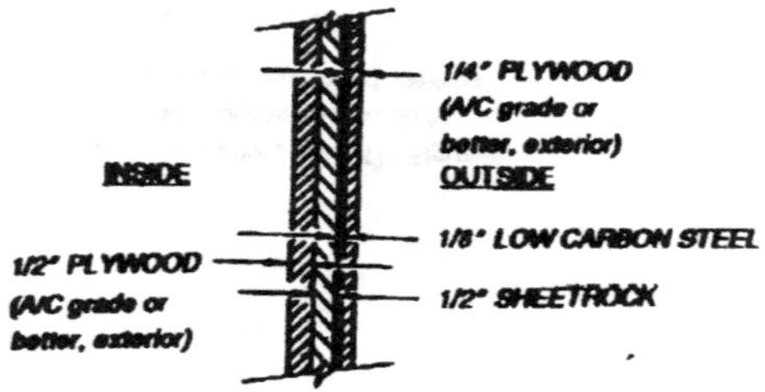

Sketch of laminate construction for container or compartment for electric blasting caps use, as illustrated in Appendix A, B, and E.

APPENDIX D

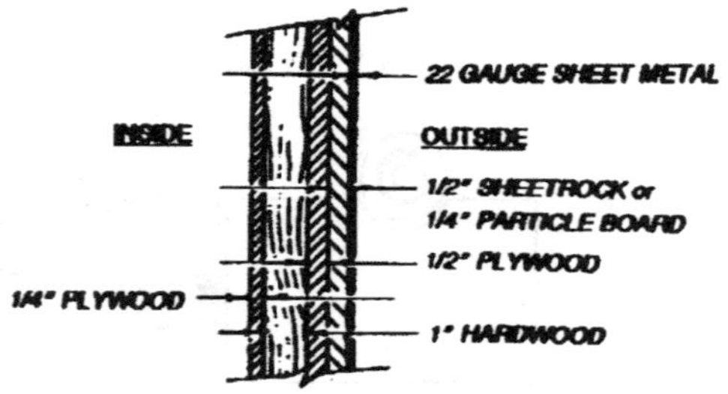

Sketch of laminate construction for container or compartment for electric blasting caps; restricted to use as illustrated in Appendix A.

[Diagrams: Courtesy of IME]

APPENDIX E

PORTABLE WHEELED TRAILERS

Figure 1

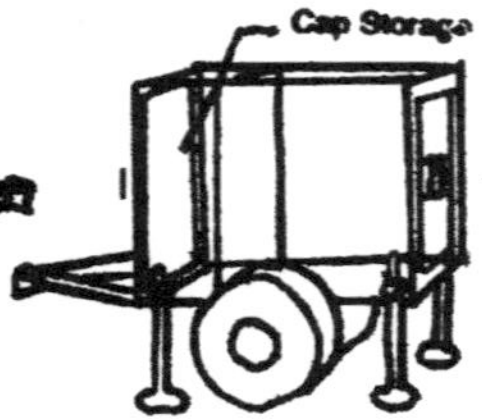

Figure 3

Figure 2

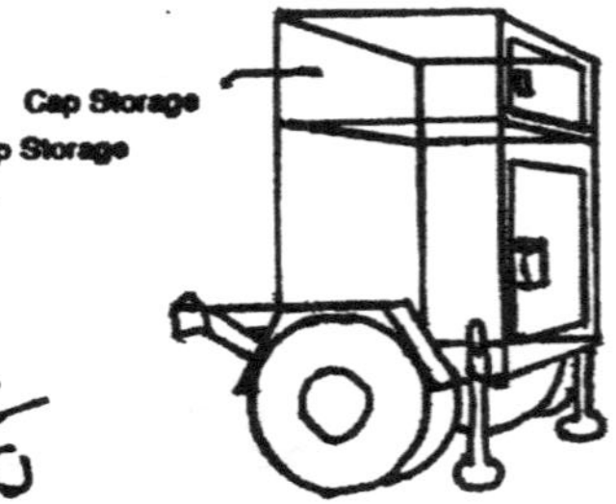

Figure 4

General Information (Revised 8/06)

Effect of 18 U.S.C. Chapter 40 On the Fireworks Industry

[Caution! This item discusses Federal requirements only. Please contact your State or local authorities for any additional requirements.]

Title XI of the Organized Crime Control Act of 1970 (18 U.S.C. Chapter 40) establishes controls over explosive materials, including black powder and other pyrotechnic compositions commonly used in fireworks. Part 555 of Title 27, Code of Federal Regulations (CFR), contains the regulations which implement Title XI. Section 555.141(a)(7) exempts "the importation, distribution, and storage of fireworks classified as UN0336, UN0337, UN0431, or UN0432 explosives by the U.S. Department of Transportation at 49 CFR 172.101 and generally known as 'consumer fireworks' or 'articles pyrotechnic.'" Section 555.141(a)(7) does not exempt "display fireworks," as defined in 555.11.

With Respect to Fireworks: Who needs a license?

1. Manufacturers of black powder;
2. Manufacturers of any other explosive material used in manufacturing consumer fireworks or display fireworks; and
3. Importers of, or dealers in, display fireworks.

With Respect to Fireworks: Who needs a permit?

1. All persons transporting, shipping, causing to be transported, or receiving display fireworks, regardless of whether for their own use or for commercial display purposes (Certain exemptions apply, e.g. agencies of the United States or of any State or political subdivisions thereof are exempt from permit requirements); and
2. A person, other than a licensee, transporting, shipping, causing to be transported, or receiving explosive materials for use in manufacturing display fireworks or consumer fireworks.

With Respect to Fireworks: Who may not need a license or permit?

Frequently, persons contracting for display fireworks (e.g., for Fourth of July observances) from a Federal explosives licensee or permittee receive a total service, including the services of a pyrotechnician who transports display fireworks in interstate or intrastate commerce to the site of the display and conducts and supervises the display. In these instances, the customers purchase and receive the contractor's services and not the explosive materials themselves (i.e. the cost of the services includes the contractor's expense in providing the fireworks utilized), and the cost of the services includes the dealer's expense in providing the fireworks utilized. When business is transacted in this manner, the customers purchasing and receiving the services need not obtain Federal explosives licenses or permits under Part 555 as long as they are not transporting, shipping, causing to be transported, or receiving explosive materials. Note: the transportation of explosive materials to the display sites would be authorized by the Federal explosives license or permit of the licensee or permittee providing the services.

With Respect to Fireworks: Types of permits

1. User permit: Allows the permit holder to transport, ship, cause to be transported, and receive display fireworks in interstate or foreign commerce for his or her own use and not for resale. This permit is issued at a cost of $100 for a 3-year period and is renewable at a cost of $50 for a 3-year period.
2. User-limited permit: Identical to the user permit but issued for a single purchase transaction, only. The fee is $75; the permit is nonrenewable.

With Respect to Fireworks: Storage

The law prohibits any person from storing any explosive materials in a manner not in conformity with the regulations promulgated by the Attorney General (18 U.S.C. 842(j)). Pursuant to this section, the Attorney General has prescribed storage regulations in 27 CFR Part 555, Subpart K. Display fireworks must be stored in conformity with the regulations. Display fireworks generally contain perchlorate mixture explosives, potassium chlorate base explosive mixtures, and black powder, which are entered on the List of Explosive Materials with numerous others. (The List, which is not all-inclusive, is annually compiled and readily available without charge from the address set out in 27 CFR 555.23 or online at www.atf.gov.) Display fireworks must be stored as low explosives in magazines meeting, at a minimum, the requirements for type 4 storage magazines prescribed by 27 CFR 555.210 unless they contain other classes of explosives. Bulk salutes must be stored as high explosives in type 1 or type 2 magazines. The net weight of the explosive materials contained in the display fireworks may be used in determining compliance with table of distance requirements. To determine the actual weight of the materials, it may be necessary to contact their manufacturers. The manufacturer of exempt or nonexempt fireworks having stocks of explosive materials on hand to be used in the manufacture of fireworks must store the stocks in conformity with applicable storage requirements.

Explosives Dealer's and User's Guide to Federal Explosives Regulation

Explosives May Not be Distributed by Licensees (Or by Any Person) to Any Person Who:

1. Is under indictment for, or who has been convicted of a crime punishable by imprisonment for a term exceeding one year.

2. Is an unlawful user of, or addicted to, marijuana or any depressant or stimulant drug or narcotic drug (as these terms are defined in section 102 of the Controlled Substances Act).

3. Has been adjudicated as a mental defective or has been committed to a mental institution.

4. Is a fugitive from justice.

5. Is an alien (with certain exceptions)

6. Has been discharged from the armed forces under dishonorable conditions; or

7. Having been a citizen of the United States, has renounced citizenship.

8. Is less than 21 years of age.

Dealers in Explosives Must:

Have a current and valid Federal explosives license.

Have proper storage facilities.

Keep accurate and complete records.

Verify that each buyer has a Federal explosives license or permit.

Verify buyers' identities.

Users of Explosives

Federal permits are required of those who transport, ship, cause to be transported, or receive explosive materials. Among other things, the permittee must keep complete and accurate records of the acquisitions and dispositions of explosives materials. Unless otherwise exempted by law, no person may receive or transport any explosive materials without a permit.

No person shall store any explosive material in a matter not in conformity with applicable regulations.

All persons must report to ATF and local authorities any loss or theft of their explosive materials within 24 hours of discovery.

A Federal license or permit does not confer any right or privilege to violate any state law or local ordinance.

The above summary is general and does not purport to fully convey the Federal explosives law and regulations pertaining to dealers and users.

Black Powder Transactions

Public Law 93-639 (1975) allows nonlicensees/nonpermittees to purchase commercially manufactured black powder, in quantities of 50 pounds or less, solely for sporting, recreational or cultural purposes for use in antique firearms or antique devices. A nonlicensee or nonpermittee purchasing black powder under the exemption need not be a resident of the State in which the dealer is located. Also, the categories of persons to whom the distribution of explosive materials is prohibited do not apply to black powder transactions made under the exemption. Acquisitions of black powder not qualifying under this exemption are subject to the same regulatory requirements that govern any other low explosive.

All persons who distribute black powder, regardless of quantity, must be licensed as explosives dealers and, among other things, must provide adequate storage.

Explosives Security

Through prompt reporting of losses and thefts of explosives and increased emphasis on physical security, explosives licensees and permittees can contribute greatly to efforts by Federal, State and local authorities to reduce the incidence of bombings and other criminal misuse of explosives in the United States. The following actions are of prime importance and in some instances required:

Report . . . any thefts or losses of explosives within 24 hours of discovery, by telephone, to ATF (toll free: 1-800-800-3855) and to appropriate local authorities. Because the States and many municipalities have designated specific agencies to investigate the theft or loss of explosives, licensees and permittees are urged to be familiar with State and local reporting procedures and appropriate contact points.

Follow . . . telephone notification with a written report on ATF Form 5400.5, "Report of Theft or Loss—Explosive Materials," to the nearest ATF Division Office, and in accordance with the instructions on the form.

Observe . . . activity around magazines, within business premises, and on job sites, particularly if strangers appear to be loitering in the area in which explosives are being kept. On-site users should take special care to assure that explosives removed from storage for use on the job are either detonated or accounted for and unused items returned to storage.

Review . . . recordkeeping practices to assure that no discrepancies exist and that no figures in reported inventories have been manipulated, and correct any clerical errors promptly. Should any questions arise concerning explosives security procedures or any aspect of explosives regulation coming under the jurisdiction of ATF, do not hesitate to contact ATF.

Note: For Q&A's on regulatory requirements governing recordkeeping and storage, see "Questions and Answers" numbers 62-87.

Additional Information

The flow of useful information is an essential ingredient in the effective administration of regulatory programs. The Bureau of Alcohol, Tobacco, Firearms and Explosives is the Federal agency charged with the responsibility of administering laws impacting the explosives industry. We call your attention to the following publication distributed by ATF that merits your attention:

The Explosives Newsletter

During 1989 ATF developed the Explosives Newsletter, an information service for Federal explosives licensees and permittees which is intended to help explosives industry members better understand the Federal laws under which they must operate. It also includes other items of particular interest to the explosives industry. There is no charge for the Explosives Newsletter; licensees and permittees automatically receive copies when new editions are published.

Explosives industry members having questions on the Federal explosives laws and regulations may address their inquiries to:

Bureau of Alcohol, Tobacco, Firearms and Explosives
Explosives Industry Programs Branch
99 New York Avenue, N.E.
Mailstop 6E405
Washington, DC 20226

Direct e-mail inquiries on general questions or variance requests may be sent to the branch at EIPB@atf.gov.

List of Explosive Materials

Pursuant to the provisions of section 841(d) of title 18, U.S.C., and 27 CFR 555.23, the Director, Bureau of Alcohol, Tobacco, Firearms and Explosives, must revise and publish in the Federal Register at least annually a list of explosives determined to be within the coverage of 18 U.S.C. Chapter 40, Importation, Manufacture, Distribution and Storage of Explosive Materials. This chapter covers not only explosives, but also blasting agents and detonators, all of which are defined as explosive materials in section 841(c) of title 18, U.S.C. Accordingly, the following is the current List of Explosive Materials subject to regulation under 18 U.S.C. Chapter 40. Materials constituting blasting agents are marked by an asterisk. While the list is comprehensive, it is not all-inclusive. The fact that an explosive material may not be on the list does not mean that it is not within the coverage definitions in section 841 of title 18, U.S.C. Explosive materials are listed alphabetically by their common names, followed by chemical names and synonyms in brackets. This revised list is effective as of September 18, 2006.

List of Explosive Materials

A

Acetylides of heavy metals.
Aluminum containing polymeric propellant.
Aluminum ophorite explosive.
Amatex.
Amatol.
Ammonal.
Ammonium nitrate explosive mixtures (cap sensitive).
*Ammonium nitrate explosive mixtures (non-cap sensitive).
Ammonium perchlorate having particle size less than 15 microns.
Ammonium perchlorate composite propellant.
Ammonium perchlorate explosive mixtures.
Ammonium picrate [picrate of ammonia, Explosive D].
Ammonium salt lattice with isomorphously substituted inorganic salts.
*ANFO [ammonium nitrate-fuel oil].
Aromatic nitro-compound explosive mixtures.
Azide explosives.

B

Baranol.
Baratol.
BEAF [1,2-bis (2, 2-difluoro-2-nitroacetoxyethane)].
Black powder.
Black powder based explosive mixtures.
*Blasting agents, nitro-carbo-nitrates, including non-cap sensitive slurry and water gel explosives.
Blasting caps.
Blasting gelatin.
Blasting powder.
BTNEC [bis (trinitroethyl) carbonate].
BTNEN [bis (trinitroethyl) nitramine].
BTTN [1,2,4 butanetriol trinitrate].
Bulk salutes.
Butyl tetryl.

C

Calcium nitrate explosive mixture.
Cellulose hexanitrate explosive mixture.
Chlorate explosive mixtures.
Composition A and variations.
Composition B and variations.
Composition C and variations.
Copper acetylide.
Cyanuric triazide.
Cyclonite [RDX].
Cyclotetramethylenetetranitramine [HMX].
Cyclotol.
Cyclotrimethylenetrinitramine [RDX].

D

DATB [diaminotrinitrobenzene].
DDNP [diazodinitrophenol].
DEGDN [diethyleneglycol dinitrate].
Detonating cord.
Detonators.
Dimethylol dimethyl methane dinitrate composition.
Dinitroethyleneurea.
Dinitroglycerine [glycerol dinitrate].
Dinitrophenol.
Dinitrophenolates.
Dinitrophenyl hydrazine.
Dinitroresorcinol.
Dinitrotoluene-sodium nitrate explosive mixtures.
DIPAM [dipicramide; diaminohexanitrobiphenyl].
Dipicryl sulfone.
Dipicrylamine.
Display fireworks.
DNPA [2,2-dinitropropyl acrylate].
DNPD [dinitropentano nitrile].
Dynamite.

E

EDDN [ethylene diamine dinitrate].
EDNA [ethylenedinitramine].
Ednatol.
EDNP [ethyl 4,4-dinitropentanoate].
EGDN [ethylene glycol dinitrate].
Erythritol tetranitrate explosives.
Esters of nitro-substituted alcohols.
Ethyl-tetryl.
Explosive conitrates.
Explosive gelatins.
Explosive liquids.
Explosive mixtures containing oxygen-releasing inorganic salts and hydrocarbons.
Explosive mixtures containing oxygen-releasing inorganic salts and nitro bodies.
Explosive mixtures containing oxygen-releasing inorganic salts and water insoluble fuels.
Explosive mixtures containing oxygen-releasing inorganic salts and water soluble fuels.
Explosive mixtures containing sensitized nitromethane.

Explosive mixtures containing tetranitromethane (nitroform).
Explosive nitro compounds of aromatic hydrocarbons.
Explosive organic nitrate mixtures.
Explosive powders.

F

Flash powder.
Fulminate of mercury.
Fulminate of silver.
Fulminating gold.
Fulminating mercury.
Fulminating platinum.
Fulminating silver.

G

Gelatinized nitrocellulose.
Gem-dinitro aliphatic explosive mixtures.
Guanyl nitrosamino guanyl tetrazene.
Guanyl nitrosamino guanylidene hydrazine.
Guncotton.

H

Heavy metal azides.
Hexanite.
Hexanitrodiphenylamine.
Hexanitrostilbene.
Hexogen [RDX].
Hexogene or octogene and a nitrated Nmethylaniline.
Hexolites.
HMTD [hexamethylenetriperoxidediamine].
HMX [cyclo-1,3,5,7-tetramethylene 2,4,6,8-tetranitramine; Octogen].
Hydrazinium nitrate/hydrazine/aluminum explosive system.
Hydrazoic acid.

I

Igniter cord.
Igniters.
Initiating tube systems.

K

KDNBF [potassium dinitrobenzofuroxane].

L

Lead azide.
Lead mannite.
Lead mononitroresorcinate.
Lead picrate.
Lead salts, explosive.
Lead styphnate [styphnate of lead, lead trinitroresorcinate].
Liquid nitrated polyol and trimethylolethane.
Liquid oxygen explosives.

M

Magnesium ophorite explosives.
Mannitol hexanitrate.
MDNP [methyl 4,4-dinitropentanoate].
MEAN [monoethanolamine nitrate].
Mercuric fulminate.
Mercury oxalate.
Mercury tartrate.
Metriol trinitrate.
Minol-2 [40% TNT, 40% ammonium nitrate, 20% aluminum].
MMAN [monomethylamine nitrate]; methylamine nitrate.
Mononitrotoluene-nitroglycerin mixture.
Monopropellants.

N

NIBTN [nitroisobutametriol trinitrate].
Nitrate explosive mixtures.
Nitrate sensitized with gelled nitroparaffin.
Nitrated carbohydrate explosive.
Nitrated glucoside explosive.
Nitrated polyhydric alcohol explosives.
Nitric acid and a nitro aromatic compound explosive.
Nitric acid and carboxylic fuel explosive.
Nitric acid explosive mixtures.
Nitro aromatic explosive mixtures.
Nitro compounds of furane explosive mixtures.
Nitrocellulose explosive.
Nitroderivative of urea explosive mixture.
Nitrogelatin explosive.
Nitrogen trichloride.
Nitrogen tri-iodide.
Nitroglycerine [NG, RNG, nitro, glyceryl trinitrate, trinitroglycerine].
Nitroglycide.
Nitroglycol [ethylene glycol dinitrate, EGDN].
Nitroguanidine explosives.
Nitronium perchlorate propellant mixtures.
Nitroparaffins Explosive Grade and ammonium nitrate mixtures.
Nitrostarch.
Nitro-substituted carboxylic acids.
Nitrourea.

O

Octogen [HMX].
Octol [75 percent HMX, 25 percent TNT].
Organic amine nitrates.
Organic nitramines.

P

PBX [plastic bonded explosives].
Pellet powder.
Penthrinite composition.
Pentolite.
Perchlorate explosive mixtures.
Peroxide based explosive mixtures.
PETN [nitropentaerythrite, pentaerythrite tetranitrate, pentaerythritol tetranitrate].
Picramic acid and its salts.
Picramide.
Picrate explosives.
Picrate of potassium explosive mixtures.
Picratol.
Picric acid (manufactured as an explosive).
Picryl chloride.

Picryl fluoride.
PLX [95% nitromethane, 5% ethylenediamine].
Polynitro aliphatic compounds.
Polyolpolynitrate-nitrocellulose explosive gels.
Potassium chlorate and lead sulfocyanate explosive.
Potassium nitrate explosive mixtures.
Potassium nitroaminotetrazole.
Pyrotechnic compositions.
PYX [2,6-bis(picrylamino)] 3,5-dinitropyridine.

R

RDX [cyclonite, hexogen, T4, cyclo-1,3,5,-trimethylene-2,4,6,-trinitramine; hexahydro-1,3,5-trinitro-S-triazine].

S

Safety fuse.
Salts of organic amino sulfonic acid explosive mixture.
Salutes (bulk).
Silver acetylide.
Silver azide.
Silver fulminate.
Silver oxalate explosive mixtures.
Silver styphnate.
Silver tartrate explosive mixtures.
Silver tetrazene.
Slurried explosive mixtures of water, inorganic oxidizing salt, gelling agent, fuel, and sensitizer (cap sensitive).
Smokeless powder.
Sodatol.
Sodium amatol.
Sodium azide explosive mixture.
Sodium dinitro-ortho-cresolate.
Sodium nitrate explosive mixtures.
Sodium nitrate-potassium nitrate explosive mixture.
Sodium picramate.
Special fireworks.
Squibs.
Styphnic acid explosives.

T

Tacot [tetranitro-2,3,5,6-dibenzo-1,3a,4,6a tetrazapentalene].
TATB [triaminotrinitrobenzene].
TATP [triacetonetriperoxide].
TEGDN [triethylene glycol dinitrate].
Tetranitrocarbazole.
Tetrazene [tetracene, tetrazine, 1(5-tetrazolyl)-4-guanyl tetrazene hydrate].
Tetrazole explosives.
Tetryl [2,4,6 tetranitro-N-methylaniline].
Tetrytol.
Thickened inorganic oxidizer salt slurried explosive mixture.
TMETN [trimethylolethane trinitrate].
TNEF [trinitroethyl formal].
TNEOC [trinitroethylorthocarbonate].
TNEOF [trinitroethylorthoformate].
TNT [trinitrotoluene, trotyl, trilite, triton].
Torpex.
Tridite.
Trimethylol ethyl methane trinitrate composition.
Trimethylolthane trinitratenitrocellulose.
Trimonite.
Trinitroanisole.
Trinitrobenzene.
Trinitrobenzoic acid.
Trinitrocresol.
Trinitro-meta-cresol.
Trinitronaphthalene.
Trinitrophenetol.
Trinitrophloroglucinol.
Trinitroresorcinol.
Tritonal.

U

Urea nitrate.

W

Water-bearing explosives having salts of oxidizing acids and nitrogen bases, sulfates, or sulfamates (cap sensitive).
Water-in-oil emulsion explosive compositions.

X

Xanthamonas hydrophilic colloid explosive mixture.

Approved: September 18, 2006.
Michael J. Sullivan, Acting Director.

[FR Doc. E6–15850 Filed 9–26–06; 8:45 am]

ATF Firearms Curios or Relics List

TABLE OF CONTENTS

COMPILER'S NOTES:

1. "*" indicates an entry made since ATF P 5300.11 was last published by way of hardcopy (10-95), supplemental, or Internet posting;
2. "S/N" and "S/Ns" mean "serial number" and "serial numbers;"
3. "Cal." and "cals." mean "caliber" and "calibers;"
4. "Mfd." means "manufactured."

This publication is a cumulative digest of determinations made by the Bureau of Alcohol, Tobacco, Firearms and Explosives (ATF) and is not inclusive of all weapons meeting curio or relic classification (i.e., firearms manufactured at least 50 years prior to the current date).

The "Introduction" section, under "Regulations Applicable to Licensed Collectors," refers to selected regulation citations applicable to licensed collector activities. The citations listed are illustrative and are meant to emphasize current regulations affecting licensed collectors. There is no intent to supercede law and regulation as described in The Gun Control Act of 1968 (GCA) as it is amended. Updated versions are published in ATF P 5300.4, Federal Firearms Regulations Reference Guide[1]. Questions arising from these citations should refer to this publication.

---

[1] ATF P 5300.4, Federal Firearms Regulations Reference Guide, can be obtained from the ATF Distribution Center (703) 455-7801.

INTRODUCTION

FEDERALLY LICENSED FIREARMS COLLECTORS

ATF is responsible for enforcing the provisions of the GCA and its subsequent amendments. A significant part of the GCA concerns the licensing and recordkeeping requirements pertaining to the manufacture, importation, distribution, and sale of firearms. A collector of Curios or Relics may obtain a collector's license under the GCA (see 18 U.S.C. Chapter 44 and the regulations issued thereunder in 27 CFR Part 478). The privileges conferred by this license extend only to transactions involving weapons classified as curio or relic firearms. In transactions involving firearms not classified as curios or relics, the licensed collector has the same status as a nonlicensee. A person need not be federally licensed to collect curios or relics. Generally, persons must be licensed in order to lawfully receive curios or relics from outside their State of residence.

The principal advantage of a Collector's License, therefore, is that a collector can acquire curios or relics in interstate commerce. Although a licensed collector may acquire and dispose of curios or relics at any location, dispositions to nonlicensees must generally be made to residents of the same State in which the collector is licensed. Further, **A LICENSED COLLECTOR IS NOT AUTHORIZED TO ENGAGE IN BUSINESS AS A DEALER IN ANY FIREARMS, INCLUDING CURIOS OR RELICS. A FEDERAL FIREARMS DEALER'S LICENSE IS REQUIRED FOR THIS ACTIVITY.** The term "engaged in business" as applied to a dealer in firearms refers, in part, to a person who devotes time, attention, and labor to engaging in such activity as a regular course of trade or business with the principal objective of livelihood and profit. Therefore, any person intending to "engage in the business" of selling firearms, including firearms defined by ATF as Curios or Relics, must first obtain a dealer's license. For example, if a collector acquires curios or relics for the purpose of sale rather than to enhance a collection, the collector would have to be licensed as a dealer in firearms under the GCA. **The sole intent and purpose of the collector's license is to enable a firearms collector to obtain curio or relic firearms from outside his or her State of residence.**

In order to obtain a collector's license, applicants need to prepare and submit ATF Form 7CR, Application for License (Collector of Curios or Relics), with the appropriate fee, $30 for 3 years, and in accordance with the instructions on the form. This form may be obtained from the Firearms and Explosives Licensing Center in Atlanta, Georgia, (404) 679-5040 or from your local ATF office.

WHAT ARE CURIOS OR RELICS?

As set out in the regulations (27 CFR 478.11), curios or relics include firearms which are of special interest to collectors by reason of some quality other than is associated with firearms intended for sporting use or as offensive or defensive weapons. To be recognized as curios or relics, firearms must fall within one of the following categories:

1.) Firearms which were manufactured at least 50 years prior to the current date, but not including replicas thereof;
2.) Firearms which are certified by the curator of a municipal, State, or Federal museum which exhibits firearms to be curios or relics of museum interest; and
3.) Any other firearms which derive a substantial part of their monetary value from the fact that they are novel, rare, bizarre, or because of their association with some historical figure, period, or event. Proof of qualification of a particular firearm under this category may be established by evidence of present value and evidence that like firearms are not available except as collector's items, or that the value of like firearms available in ordinary commercial channels is substantially less.

SELECTED REGULATIONS (27 C.F.R.) APPLICABLE TO LICENSED COLLECTORS

§ 478.11 Meaning of terms. Collector: Any person who acquires, holds, or disposes of firearms as curios or relics.

Collection premises: The premises described on the license of a collector as the location at which he maintains his collection of curios and relics. **Licensed collector:** A collector of curios and relics only and licensed under the provisions of this part. **Person:** Any individual, corporation, company, association, firm, partnership, society, or joint stock company.

§ 478.23 **Right of entry and examination.** (a) [A]ny ATF officer, when there is reasonable cause to believe a violation of the Act has occurred and that evidence of the violation may be found on the premises of any licensed manufacturer, licensed importer, licensed dealer, or licensed collector, may, upon demonstrating such cause before a Federal magistrate and obtaining from the magistrate a warrant authorizing entry, enter during business hours (or, in the case of a licensed collector, the hours of operation) the premises, including places of storage, of any such licensee for the purpose of inspecting or examining:

1) Any records or documents required to be kept by such licensee under this part and
2) Any inventory of firearms or ammunition kept or stored by any licensed manufacturer, licensed importer, or licensed dealer at such premises or any firearms curios or relics or ammunition kept or stored by any licensed collector at such premises.

(c) Any ATF officer, without having reasonable cause to believe a violation of the Act has occurred or that evidence of the violation may be found and without demonstrating such cause before a Federal magistrate or obtaining from the magistrate a warrant authorizing entry, may enter during hours of operation the premises, including places of storage, of any licensed collector for the purpose of inspecting or examining the records, documents, firearms, and ammunition referred to in paragraph (a) of this section (1) for ensuring compliance with the recordkeeping requirements of this part not more than once during any 12-month period or (2) when such inspection or examination may be required for determining the disposition of one or more particular firearms in the course of a bona fide criminal investigation. At the election of the licensed collector, the annual inspection permitted by this paragraph shall be performed at the ATF office responsible for conducting such inspection in the closest proximity to the collector's premises.

(d) The inspections and examinations provided by this section do not authorize an ATF officer to seize any records or documents other than those records or documents constituting material evidence of a violation of law. If an ATF officer seizes such records or documents, copies shall be provided the licensee within a reasonable time.

§ **478.25a Responses to requests for information.** Each licensee shall respond immediately to, and in no event later than 24 hours after the receipt of, a request by an ATF officer at the National Tracing Center for information contained in the records required to be kept by this part for determining the disposition of one or more firearms in the course of a bona fide criminal investigation. The requested information shall be provided orally to the ATF officer within the 24-hour period. Verification of the identity and employment of National Tracing Center personnel requesting information may be established at the time the requested information is provided by telephoning the toll-free number 1-800-788-7133 or using the toll-free facsimile (FAX) number 1-800-578-7223.

§ **478.29 Out-of-State acquisition of firearms by nonlicensees.** No person, other than a licensed importer, licensed manufacturer, licensed dealer, or licensed collector, shall transport into or receive in the State where the person resides (or if a corporation or other business entity, where it maintains a place of business) any firearm purchased or otherwise obtained by such person outside that State: **Provided,** That the provisions of this section:

(a) Shall not preclude any person who lawfully acquires a firearm by bequest or intestate succession in a State other than his State of residence from transporting the firearm into or receiving it in that State, if it is lawful for such person to purchase or possess such firearm in that State,
(b) Shall not apply to the transportation or receipt of a rifle or shotgun obtained from a licensed manufacturer, licensed importer, licensed dealer, or licensed collector in a State other than the transferee's State of residence in an over-the-counter transaction at the licensee's premises obtained in conformity with the provisions of § 478.96(c), and
(c) Shall not apply to the transportation or receipt of a firearm obtained in conformity with the provisions of §§ 478.30 and 478.97.

§ **478.29a Acquisition of firearms by nonresidents.** No person, other than a licensed importer, licensed manufacturer, licensed dealer, or licensed collector, who does not reside in any State shall receive any firearms unless such receipt is for lawful sporting purposes.

§ **478.31 Delivery by common or contract carrier.** (a) No person shall knowingly deliver or cause to be delivered to any common or contract carrier for transportation or shipment in interstate or foreign commerce to any person other than a licensed importer, licensed manufacturer, licensed dealer, or licensed collector, any package or other container in which there is any firearm or ammunition without written notice to the carrier that such firearm or ammunition is being transported or shipped:

Provided, That any passenger who owns or legally possesses a firearm or ammunition being transported aboard any common or contract carrier for movement with the passenger in interstate or foreign commerce may deliver said firearm or ammunition into the custody of the pilot, captain, conductor or operator of such common or contract carrier for the duration of that trip without violating any provision of this part.

(b) No common or contract carrier shall require or cause any label, tag, or other written notice to be placed on the outside of any package, luggage, or other container indicating that such package, luggage, or other container contains a firearm.

(c) No common or contract carrier shall transport or deliver in interstate or foreign commerce any firearm or ammunition with knowledge or reasonable cause to believe that the shipment, transportation, or receipt thereof would be in violation of any provision of this part:

Provided, however, That the provisions of this paragraph shall not apply in respect to the transportation of firearms or ammunition in in-bond shipment under Customs laws and regulation.

(d) No common or contract carrier shall knowingly deliver in interstate or foreign commerce any firearm without obtaining written acknowledgement of receipt from the recipient of the package or other container in which there is a firearm: **Provided,** That this paragraph shall not apply with respect to the return of a firearm to a passenger who places firearms in the carrier's custody for the duration of the trip.

§ 478.39a Reporting theft or loss of firearms. Each licensee shall report the theft or loss of a firearm from the licensee's inventory (including any firearm which has been transferred from the licensee's inventory to a personal collection and held as a personal firearm for at least 1 year), or from the collection of a licensed collector, within 48 hours after the theft or loss is discovered. Licensees shall report thefts or losses by telephoning 1-800-800-3855 (nationwide toll-free number) and by preparing ATF Form 3310.11, Federal Firearms Licensee Theft/Loss Report[2], in accordance with the instructions on the form. The original of the report shall be forwarded to the office specified thereon, and Copy 1 shall be retained by the licensee as part of the licensee's permanent records. Theft or loss of any firearm shall also be reported to the appropriate local authorities.

§ 478.41 General. (c) Each person seeking the privileges of a collector licensed under this part shall file an application, with the required fee (see § 478.42), with ATF in accordance with the instructions on the form (see § 478.44), and pursuant to § 478.47, receive from the Chief, National Licensing Center, the license covering the collection of curios and relics. A separate license may be obtained for each collection premises, and such license shall, subject to the provisions of the Act and other applicable provisions of law, entitle the licensee to transport, ship, receive, and acquire curios and relics in interstate or foreign commerce, and to make disposition of curios and relics in interstate or foreign commerce, to any other person licensed under provisions of this part, for the period stated on the license.

(d) The collector license provided by this part shall apply only to transactions related to a collector's activity in acquiring, holding, or disposing of curios and relics. A collector's license does not authorize the collector to engage in a business required to be licensed under the Act or this part. Therefore, if the acquisitions and dispositions of curios and relics by a collector bring the collector within the definition of a manufacturer, importer, or dealer under this part, he shall qualify as such.

§ 478.44 Original license. (b) Any person who desires to obtain a license as a collector under the Act and this part, or who has not timely submitted an application for renewal of the previous license issued under this part, shall file an application, ATF Form 7CR (Curios and Relics), with ATF in accordance with the instructions on the form. The application must be executed under the penalties of perjury and the penalties imposed by 18 U.S.C. 924. The application shall be accompanied by a completed ATF Form 5300.37 and ATF Form 5300.36 and shall include the appropriate fee

[2] ATF F 3310.11, Federal Firearms Licensee Theft/Loss Report, can be obtained from the ATF Distribution Center, (703) 455-7801.

in the form of a money order or check made payable to the Bureau of Alcohol, Tobacco, Firearms and Explosives. ATF Forms 7CR (Curios and Relics), ATF Forms 5300.37, and ATF Forms 5300.36 may be obtained by contacting any ATF office.

§ 478.47 Issuance of license. (a) Upon receipt of a properly executed application for a license on ATF Form 7, ATF Form 7CR, or ATF Form 8 Part II, the Chief, National Licensing Center, shall, upon finding through further inquiry or investigation, or otherwise, that the applicant is qualified, issue the appropriate license. Each license shall bear a serial number, and such number may be assigned to the licensee to whom issued for so long as the licensee maintains continuity of renewal in the same location (State).

(b) The Chief, National Licensing Center, shall approve a properly executed application for license on ATF Form 7, ATF Form 7CR, or ATF Form 8 Part II, if:

(1) The applicant is 21 years of age or over;

(2) The applicant (including, in the case of a corporation, partnership, or association, any individual possessing, directly or indirectly, the power to direct or cause the direction of the management and policies of the corporation, partnership, or association) is not prohibited under the provisions of the Act from shipping or transporting in interstate or foreign commerce, or possessing in or affecting commerce, any firearm or ammunition, or from receiving any firearm or ammunition which has been shipped or transported in interstate or foreign commerce;

(3) The applicant has not willfully violated any of the provisions of the Act or this part;

(4) The applicant has not willfully failed to disclose any material information required, or has not made any false statements as to any material fact, in connection with his application;

(5) The applicant has in a State

  i.) premises from which he conducts business subject to license under the Act or from which he intends to conduct such business within a reasonable period of time, or in the case of a collector, premises from which he conducts his collecting subject to license under the Act or from which he intends to conduct such collecting within a reasonable period of time; and

(6) The applicant has filed an ATF Form 5300.37 (Certification of Compliance with State and Local Law) with ATF in accordance with the instructions on the form certifying under penalties of perjury that –

  (i) The business to be conducted under the license is not prohibited by State or local law in the place where the licensed premises are located;

  (ii) Within 30 days after the application is approved the business will comply with the requirements of State and local law applicable to the conduct of business;

  (iii) The business will not be conducted under the license until the requirements of State and local law applicable to the business have been met; and

  (iv) The applicant has completed and sent or delivered ATF F 5300.36 (Notification of Intent to Apply for a Federal Firearms License) to the chief law enforcement officer of the locality in which the premises are located, which indicates that the applicant intends to apply for a Federal firearms license. For purposes of this paragraph, the "chief law enforcement officer" is the chief of police, the sheriff, or an equivalent officer.

(c) The Chief, National Licensing Center, shall approve or the Director of Industry Operations shall deny an application for license within the 60-day period beginning on the date the properly executed application was received: **Provided,** That when an applicant for license renewal is a person who is, pursuant to the provisions of § 478.78, § 478.143, or § 478.144, conducting business or collecting activity under a previously issued license, action regarding the application will be held in abeyance pending the completion of the proceedings against applicant's existing license or license application, final determination of the applicant's criminal case, or final action by the Director on an application for relief submitted pursuant to § 478.144, as the case may be.

(d) When the Director of Industry Operations or the Chief, National Licensing Center, fails to act on an application for license within the 60-day period prescribed by paragraph (c) of this section, the applicant may file an action under section 1361 of title 28, United States Code, to compel ATF to act upon the application.

§ 478.49 Duration of license. The license entitles the person to whom issued to engage in the business or activity specified on the license, within the limitations of the Act and the regulations contained in this part, for a three year period, unless terminated sooner.

§ 478.50 Locations covered by license. The license covers the class of business or the activity specified in the license at the address specified therein. A separate license must be obtained for each location at which a firearms or ammunition business or activity requiring a license under this part is conducted except:

(a) No license is required to cover a separate warehouse used by the licensee solely for storage of firearms or ammunition if the records required by this part are maintained at the licensed premises served by such warehouse;

(b) A licensed collector may acquire curios and relics at any location, and dispose of curios or relics to any licensee or to other persons who are residents of the State where the collector's license is held and the disposition is made.[3]

(c) A licensee may conduct business at a gun show pursuant to the provision of § 478.100; or

(d) A licensed importer, manufacturer, or dealer may engage in the business of dealing in curio or relic firearms with another licensee at any location pursuant to the provision of § 478.100.

§ 478.58 State or other law. A license issued under this part confers no right or privilege to conduct business or activity contrary to State or other law. The holder of such a license is not by reason of the rights and privileges granted by that license immune from punishment for operating a firearm or ammunition business or activity in violation of the provisions of any State or other law. Similarly, compliance with the provisions of any State or other law affords no immunity under Federal law or regulations.

§ 478.91 Posting of License. Any license issued under this part shall be kept posted and kept available for inspection on the premises covered by the license.

§ 478.94 Sales or deliveries between licensees. A licensed importer, licensed manufacturer, or licensed dealer selling or otherwise disposing of firearms, and a licensed collector selling or otherwise disposing of curios or relics, to another licensee shall verify the identity and licensed status of the transferee prior to making the transaction. Verification shall be established by the transferee furnishing to the transferor a certified copy of the transferee's license and by such other means as the transferor deems necessary: **Provided,** That it shall not be required **(a)** for a transferee who has furnished a certified copy of its license to a transferor to again furnish such certified copy to that transferor during the term of the transferee's current license **(b)** for a licensee to furnish a certified copy of its license to another licensee if a firearm is being returned either directly or through another licensee to such licensee and **(c)** for licensees of multilicensed business organizations to furnish certified copies of their licenses to other licensed locations operated by such organization: **Provided further,** That a multilicensed business organization may furnish to a transferor, in lieu of a certified copy of each license, a list, certified to be true, correct, and complete, containing the name, address, license number, and the date of license expiration of each licensed location operated by such organization, and the transferor may sell or otherwise dispose of firearms as provided by this section to any licensee appearing on such list without requiring a certified copy of a license therefrom. A transferor licensee who has the certified information required by this section may sell or dispose of firearms to a licensee for not more than 45 days following the expiration date of the transferee's license.

§ 478.96 Out-of-State and mail order sales. (c)(1) A licensed importer, licensed manufacturer, or licensed dealer may sell or deliver a rifle or shotgun, and a licensed collector may sell or deliver a rifle or shotgun that is a curio or relic to a nonlicensed resident of a State other than the State in which the licensee's place of business is located if –

- (i) The purchaser meets with the licensee in person at the licensee's premises to accomplish the transfer, sale, and delivery of the rifle or shotgun;
- (ii) The licensed importer, licensed manufacturer, or licensed dealer complies with the provisions of § 478.102;

---

[3] Exception to this provision is provided for in **§ 178.96(c)(1)(i) in which a licensed collector is specifically authorized to sell a curio or relic shotgun or rifle to a nonlicensed resident of another State as long as the transaction is consummated at the licensed collector's premises and the transaction is legal in both States.**

(iii) The purchaser furnishes to the licensed importer, licensed manufacturer, or licensed dealer the firearms transaction record, Form 4473, required by § 478.124; and
(iv) The sale, delivery, and receipt of the rifle or shotgun fully comply with the legal conditions of sale in both such States.

§ 478.99 Certain prohibited sales or deliveries. (a) Interstate sales or deliveries. A licensed importer, licensed manufacturer, licensed dealer, or a licensed collector shall not sell or deliver any firearm to any person not licensed under this part and who the licensee knows or has reasonable cause to believe does not reside in (or if a corporation or other business entity does not maintain a place of business in) the State in which the licensee's place of business or activity is located. **Provided,** That the foregoing provisions of this paragraph **(1)** shall not apply to the sale or delivery of a rifle or shotgun (curio or relic, in the case of a licensed collector) to a resident of a State other than the State in which the licensee's place of business or collection premises is located if the requirements of § 478.96(c) are fully met, and **(2)** shall not apply to the loan or rental of a firearm to any person for temporary use for lawful sporting purposes. (See § 478.97)

(b) Sales or deliveries to underage persons. A licensed importer, licensed manufacturer, licensed dealer, or licensed collector shall not sell or deliver **(1)** any firearm or ammunition to any individual who the importer, manufacturer, dealer, or collector knows or has reasonable cause to believe is less than 18 years of age, and, if the firearm, or ammunition, is other than a shotgun or rifle, or ammunition for a shotgun or rifle, to any individual who the importer, manufacturer, dealer, or collector knows or has reasonable cause to believe is less than 21 years of age, or **(2)** any firearm to any person in any State where the purchase or possession by such person of such firearm would be in violation of any State law or any published ordinance applicable at the place of sale, delivery, or other disposition, unless the importer, manufacturer, dealer, or collector knows or has reasonable cause to believe that the purchase or possession would not be in violation of such State law or such published ordinance.

(c) Sales or deliveries to prohibited categories of persons. A licensed manufacturer, licensed importer, licensed dealer, or licensed collector shall not sell or otherwise dispose of any firearm or ammunition to any person knowing or having reasonable cause to believe that such person:

(1) Is, except as provided by § 478.143, under indictment for, or, except as provided by § 478.144, has been convicted in any court of a crime punishable by imprisonment for a term exceeding 1 year;
(2) Is a fugitive from justice;
(3) Is an unlawful user of or addicted to any controlled substance (as defined in section 102 of the Controlled Substance Act, 21 U. S. C. 802);
(4) Has been adjudicated as mentally defective or has been committed to any mental institution;
(5) Is an alien illegally or unlawfully in the United States; **Provided** That the provisions of this paragraph (c)(5) do not apply to any nonimmigrant alien if that alien is –
  (i) Admitted to the United States for lawful hunting or sporting purpose or is in possession of a hunting license or permit lawfully issued in the United States;
  (ii) An official representative of a foreign government who is either accredited to the United States Government or the Government's mission to an international organization having its headquarters in the United States or en route to or from another country to which that alien is accredited. This exception only applies if the firearm or ammunition is shipped, transported, possessed, or received in the representative's official capacity;
  (iii) (iii) An official of a foreign government or a distinguished foreign visitor who has been so designated by the Department of State. This exception only applies if the firearm or ammunition is shipped, transported, possessed, or received in the official's or visitor's official capacity, except if the visitor is a private individual who does not have an official capacity; or
  (iv) A foreign law enforcement officer of a friendly foreign government entering the United States on official law enforcement business;
(6) Has been discharged from the Armed Forces under dishonorable conditions;
(7) Who, having been a citizen of the United States, has renounced citizenship;
(8) Is subject to a court order that restrains such person from harassing, stalking, or threatening an intimate partner of such person or child of such intimate partner or person, or engaging in other conduct that would place an intimate partner in reasonable fear of bodily injury to the partner or child, except that this paragraph shall only apply to a court order that
  (i) Was issued after a hearing of which such person received actual notice, and at which such person had the opportunity to participate; and

(ii) (A) Includes a finding that such person represents a credible threat to the physical safety of such intimate partner or child; or

(B) By its terms explicitly prohibits the use, attempted use, or threatened use of physical force against such intimate partner or child that would reasonably be expected to cause bodily injury,

or

(9) Has been convicted of a misdemeanor crime of domestic violence.

§ 478.103 Posting of signs and written notification to purchasers of handguns. **(a)** Each licensed importer, manufacturer, dealer, or collector who delivers a handgun to a nonlicensee shall provide such nonlicensee with written notification as described in paragraph (b) of this section.

(b) The written notification (ATF I 5300.2)[4] required by paragraph (a) of this section shall state as follows:

1.) The misuse of handguns is a leading contributor to juvenile violence and fatalities.
2.) Safely storing and securing firearms away from children will help prevent the unlawful possession of handguns by juveniles, stop accidents, and save lives.
3.) Federal law prohibits, except in certain limited circumstances,[5] anyone under 18 years of age from knowingly possessing a handgun, or any person from transferring a handgun to a person under 18.
4.) A knowing violation of the prohibition against selling, delivering, or otherwise transferring a handgun to a person under the age of 18 is, under certain circumstances, punishable by up to 10 years in prison.

§ 478.125 Record of receipt and disposition (f) Firearms receipt and disposition by licensed collectors. Each licensed collector shall enter into a record each receipt and disposition of firearms curios or relics. The record required by this paragraph shall be maintained in bound form under the format prescribed below. The purchase or other acquisition of a curio or relic shall, except as provided in [§ 27 C.F.R 478.125] (g) of this section, be recorded no later than the close of the next business day following the date of such purchase or other acquisition. The record shall show the date of receipt; the name and address or the name and license number of the person from whom received; the name of the manufacturer and importer (if any); the model, serial number, type, and the caliber or gauge of the firearm curio or relic. The sale or other disposition of a curio or relic shall be recorded by the licensed collector no later than 7 days following the date of such transaction. When such disposition is made to a licensee, the commercial record of the transaction shall be retained, until the transaction is recorded, separate from other commercial documents maintained by the licensee, and be readily available for inspection. The record shall show the date of the sale or other disposition of each firearm curio or relic, the name and address of the person to whom the firearm curio or relic is transferred, or the name and license number of the person to whom transferred if such person is a licensee, and the date of birth of the transferee if other than a licensee. In addition, the licensee shall:

1) Cause the transferee, if other than a licensee, to be identified in any manner customarily used in commercial transactions (e.g., a driver's license), and note on the record the method used, and
2) In the case of a transferee who is an alien legally in the United States and who is other than a licensee –
   (i) Verify the identity of the transferee by examining an identification document (as defined in § 478.11), and
   (ii) Cause the transferee to present documentation establishing that the transferee is a resident of the State (as defined in § 478.11) in which the licensee's business premises is located if the firearm curio or relic is other than a shotgun or rifle, and note on the record the documentation used or is a resident of any State and has resided in such State continuously for at least 90 days prior to the transfer of the firearm if the firearm curio or relic is a shotgun or rifle and shall note on the record the documentation used. Examples of acceptable documentation include utility bills or a lease agreement which show that the transferee has resided in the State continuously for at least 90 days prior to the transfer of the firearm curio or relic.
3) The format required for the record of receipt and disposition of firearms by collectors is as follows:

Firearms Collectors Acquisition and Disposition Record

Description of firearm							Disposition				
Manufacturer and/or importer	Model	Serial No.	Type	Caliber or gauge	Date	Name and address or name and license No.	Date	Name and address or name and license No.	Date of birth if nonlicensee	Driver's license No. or other identification if nonlicensee	For transfers to aliens, documentation used to establish residency

§ 478.126a **Reporting multiple sales or other disposition of pistols and revolvers.** Each licensee shall prepare a report of multiple sales or other disposition whenever the licensee sells or otherwise disposes of, at one time or during any 5 consecutive business days, two or more pistols, or revolvers, or any combination of pistols and revolvers totaling two or more, to an unlicensed person: **Provided**, That a report need not be made where pistols or revolvers, or any combination thereof, are returned to the same person from whom they were received. The report shall be prepared on Form 3310.4, Report of Multiple Sale or Other Disposition of Pistols and Revolvers[6]. Not later than the close of business on the day that the multiple sale or other disposition occurs, the licensee shall forward two copies of Form 3310.4 to the ATF office specified thereon and one copy to the State police or to the local law enforcement agency in which the sale or other disposition took place. Where the State or local law enforcement officials have notified the licensee that a particular official has been designated to receive Forms 3310.4, the licensee shall forward such forms to that designated official. The licensee shall retain one copy of Form 3310.4 and attach it to the firearms transaction record,[7] Form 4473, executed upon delivery of the pistols or revolvers.

§ **478.129 Record Retention (e) Records of dealers and collectors under the Act.** The records prepared by licensed dealers and licensed collectors under the Act of the sale or other disposition of firearms and the corresponding record of receipt of such firearms shall be retained through December 15, 1988, after which records of transactions over 20 years of age may be discarded.

§ **478.147 Return of firearm.** A person not otherwise prohibited by Federal, State, or local law may ship a firearm to a licensed importer, licensed manufacturer, or licensed dealer for any lawful purpose, and, notwithstanding any other provision of this part, the licensed manufacturer, licensed importer, or licensed dealer may return in interstate or foreign commerce to that person the firearm or a replacement firearm of the same kind and type. (See § 478.124(a) for requirements of a Form 4473 prior to return.) A person not otherwise prohibited by Federal, State, or local law may ship a firearm curio or relic to a licensed collector for any lawful purpose, and, notwithstanding any other provision of this part, the licensed collector may return in interstate or foreign commerce to the person the firearm curio or relic.

CURIOS OR RELICS AND COLLECTOR'S ITEMS: DEFINITIONS AND DETERMINATIONS

27 CFR § 479.24 **Destructive device determination**. The Director shall determine, in accordance with 26 U.S.C. 5845(f), whether a device is excluded from the definition of a destructive device. A person who desires to obtain a determination under that provision of law for any device which he believes is not likely to be used as a weapon shall submit a written request, in triplicate, for a ruling thereon to the Director. Each such request shall be executed under the penalties of perjury and contain a complete and accurate description of the device, the name and address of the manufacturer or importer thereof, the purpose of and use for which it is intended, and such photographs, diagrams, or drawings as may be necessary to enable the Director to make his determination. The Director may require the submission to him of a sample of such device for examination and evaluation. If the submission of such device is impracticable, the person requesting the ruling shall so advise the Director and designate the place where the device will be available for examination and evaluation.

27 CFR § 479.25 **Collector's items.** The Director shall determine, in accordance with 26 U.S.C. 5845(a), whether a firearm or device, which although originally designed as a weapon, is by reason of the date of its manufacture, value, design, and other characteristics primarily a collector's item and is not likely to be used as a weapon. A person who desires to obtain a determination under that provision of law shall follow the procedures prescribed in 27 CFR § 479.24 relating to destructive device determinations, and shall include information as to date of manufacture, value, design and other characteristics which would sustain a finding that the firearm or device is primarily a collector's item and is not likely to be used as a weapon.

27 CFR § 478.26 **Curio and relic determination.** Any person who desires to obtain a determination whether a particular firearm is a curio or relic shall submit a written request, in duplicate, for a ruling thereon to the Director. Each such request shall be executed under the penalties of perjury and shall contain a complete and accurate description of the

---

[6] ATF F 3310.4, Report of Multiple Sales or Other Disposition of Pistols and Revolvers, can be obtained from the ATF Distribution Center (703) 455-7801.

[7] Licensed collectors are not required to execute ATF Form 4473 for transfers of curios or relic firearms.

firearm, and such photographs, diagrams, or drawings as may be necessary to enable the Director to make a determination. The Director may require the submission of the firearm for examination and evaluation. If the submission of the firearm is impractical, the person requesting the determination shall so advise the Director and designate the place where the firearm will be available for examination and evaluation.

COLLECTORS WANTING A DETERMINATION

Collectors desiring to obtain a determination as to whether a particular firearm qualifies for classification as a curio or relic in accordance with 27 CFR 478.11 and 478.26 or a collector's item in accordance with § 479.25 and § 479.24 should submit a written request for a ruling. The letter should include:

1) A complete physical description of the item;
2) Reasons the collector believes the item merits the classification; and
3) Data concerning the history of the item, including production figures, if available, and market value.

As stated in the regulations, submission of the firearm may be required prior to a determination being made. Please submit your request to:

Bureau of Alcohol, Tobacco, Firearms and Explosives
Firearms Technology Branch
244 Needy Road
Martinsburg, WV 25405

IMPORTATION RESTRICTIONS

Licensed collectors may lawfully import curio or relic firearms other than surplus military firearms, nonsporting firearms, and NFA weapons. [A surplus military firearm is defined as one that belonged to a regular or irregular military force at any time. Alteration of the firearm does not change its status. Therefore, a sporting firearm with a surplus military frame or receiver is a surplus military firearm, because a frame or receiver is classified as a firearm as described in 18 U.S.C. § 921(a)(3).] Surplus military firearms are generally prohibited from importation under 18 U.S.C. § 925(d)(3). However, 18 U.S.C. § 925(e) authorizes licensed importers (FFL type 08 or 11) to import surplus military rifles, shotguns, and handguns classified as curios or relics; provided that such handguns are generally recognized as particularly suitable for or readily adaptable to sporting purposes. Nonsporting handguns are those pistols and revolvers that do not meet size and safety prerequisites, or that do not accrue a qualifying score on ATF Form 4590, "Factoring Criteria for Weapons." Surplus military firearms classified as curios or relics must be in their original military configuration to qualify for importation.

NFA weapons are those "firearms" defined in § 5845(a) of the National Firearms Act (26 U.S.C. Chapter 53). These include machineguns, silencers, shotguns having a barrel less than 18 inches in length, rifles having a barrel less than 16 inches in length, smooth bore handguns, disguised weapons, certain firearms having a bore greater than one-half inch in diameter, destructive devices (e.g., bombs, grenades, and mines), and certain other firearms. A licensed collector may not import NFA firearms.

Firearms may only be imported pursuant to a Form 6 issued by ATF. Collectors having questions concerning the importability of specific curio or relic firearms should contact:

Bureau of Alcohol, Tobacco, Firearms and Explosives
Firearms and Explosives Imports Branch
244 Needy Road
Martinsburg, WV 25405
Telephone: (304) 616-4550
Facsimile: (304) 616-4551

THE FIREARMS CURIOS OR RELICS LIST
THIS LIST IS COMPOSED OF FOUR MAJOR SECTIONS:

SECTION I. **Ammunition Classified as Curios or Relics:** Ammunition is no longer classified as "curios or relics" since the Congress in 1986 removed the interstate controls over ammunition under the GCA. We have continued the listing of 1972-1986 ammunition determinations as a service to those who rely on this list as a means of evaluation.

SECTION II. **Firearms Classified as Curios or Relics Under 18 U.S.C. Chapter 44:** Licensed collectors may acquire, hold, or dispose of these firearms as curios or relics. However, they are still "firearms" as defined in 18 U.S.C. 921(a)(3) and are, therefore, subject to all Chapter 44 controls. Generally, this category includes commemorative firearms, pistols, revolvers, rifles, and shotguns.

SECTION III. **Weapons Removed From the National Firearms Act as Collectors' Items and Classified as "Curios or Relics" Under 18 U.S.C. Chapter 44:** Weapons in this section are excluded entirely from the provisions of the National Firearms Act (NFA), 26 U.S.C. Chapter 53. Thus, approval from ATF to transfer these weapons is not required. They need not be registered in the National Firearms Registration and Transfer Record, and there is no transfer tax liability. However, these weapons are still "firearms" under Chapter 44, and remain subject to regulation under Part 478. Licensed collectors may receive these weapons in both intrastate and interstate commerce, and may transfer them intrastate, to both licensees and nonlicensees, and interstate, to licensed collectors and other licensees. While transfer may be made interstate to another licensee, they cannot be shipped interstate to a nonlicensee.

NOTE: Certain antique firearms are discussed and listed in Section IIIA.

SECTION IV. **National Firearms Act Weapons Classified as "Curios or Relics" Under 18 U.S.C. Chapter 44:** These weapons (e.g., machineguns) are firearms within the scope of the National Firearms Act (NFA), 26 U.S.C. Chapter 53, and are subject to all NFA Act provisions. Accordingly, these weapons cannot be lawfully transferred or received unless they are registered in the National Firearms Registration and Transfer Record at Bureau Headquarters. Once the registration requirements have been met, transfer may be made either intrastate or interstate to licensed collectors on ATF Form 4, "Application For Tax Paid Transfer and Registration of a Firearm." In each instance, first the appropriate transfer tax must be paid; ATF may then approve the transfer application. The application to transfer must be accompanied by an individual transferee's fingerprints and photograph as prescribed on the form. The form also requires a law enforcement official's certification, which must be completed.

Unserviceable weapons are still subject to the provisions of the National Firearms Act, except that they may be transferred free of transfer tax on ATF Form 5, "Application For Tax Exempt Transfer and Registration of a Firearm." As defined in 18 U.S.C. 5845(h) a firearm is unserviceable if it "is incapable of discharging a shot … [and] of being readily restored to a firing condition." Questions concerning the lawfulness of transactions in these weapons should be directed to:

Bureau of Alcohol, Tobacco, Firearms and Explosives
National Firearms Act Branch
244 Needy Road
Martinsburg, WV 25405
Telephone: (304) 616-4500
Facsimile: (304) 616-4501

Questions concerning the criteria for rendering a specific weapon unserviceable should be directed to:

Bureau of Alcohol, Tobacco, Firearms and Explosives
Firearms Technology Branch
244 Needy Road
Martinsburg, WV 25405
Telephone: (304) 616-4468
Facsimile: (304) 616-4443

NOTE: THE FOLLOWING LIST OF CURIOS OR RELICS IS NOT ALL-INCLUSIVE.

Questions concerning classification of a particular firearm should be directed to:

Bureau of Alcohol, Tobacco, Firearms and Explosives
Firearms Technology Branch
244 Needy Road
Martinsburg, WV 25405
Telephone: (304) 616-4468
Facsimile: (304) 616-4443

* * * * *

SECTION I: Ammunition Classified As Curios Or Relics

The following ammunition was classified as curios or relics prior to August 1986.

1. FOREIGN MILITARY RIFLE CARTRIDGES

.557 x 450 cal. Martini Henry
6.5 x 54Rmm Mannlicher (Dutch and Rumanian)
6.5 x 55mm Swedish Mauser
7.62 x 54Rmm Russian
7.92mm anti-tank cartridge for the German PzB 38, PzB 39, & Gr 39 anti-tank rifle.
7.92mm Kurz (7.92 x 33mm), as originally produced prior to 1946.
8 x 50Rmm Lebel
8 x 53mm Japanese Murata
8 x 56Rmm Portuguese Kropacheck, mfd. in Austria, for the Govt. of Portugal.
9.5 x 60mm Turkish Mauser
10.15 x 63mm Serbian Mauser
10.4 x 47Rmm Italian Vefterli M70
11 x 50mm Belgian Albini M67/72
11 x 53mm Belgian Comblain M71
11 x 60 Japanese Murata
11.15 x 60Rmm (.43) Mauser
11.3 x 50mm Beaumont M71
11.4 x 50mm Brazilian Comblain M74
11.43 x 55mm Turkish
11.7 x 51 mm Danish Remington
6.5 x 52mm Italian (Mannlicher-Carcano)
6.5 x 58mm Portuguese Vergueiro
7.5 x 54mm Swiss Schmidt-Rubin
7.65mm Argentine Navy Match 7.92mm anti-tank cartridge for the Polish (Marosczek) anti-tank rifle described as a cal. 7.92mm, Gr 39 Granatbuchse.
8 x 50Rmm Austrian Mannlicher
8 x 52Rmm Siamese Mauser, mfd. in Japan, for the Government of Siam
8 x 58Rmm Danish Krag
8 x 60mm Guedes M85 Portuguese
10.15 x 61mm Jarman
10.4 x 38Rmm Swiss Vefteril M69/81
10.75 x 58mm Russian Berdan
11 x 52mm Netherlands Beaumont M71/78
11 x 59mm French Gras
11.15 x 58mm Austrian Werndl M77
11.15 x 58mm (.43) Spanish Remington
11.4 x 50mm Austrian Werndl M73
11.43 x 50mm (.43) Egyptian Remington
11.5 x 57mm (.43) Spanish Reformado
12.11 x 44mm Remington M67 (Norway & Sweden)

2. EUROPEAN SPORTING RIFLE CARTRIDGES

6 x 29.5mm Stahl
6 x 58mm Forester (Rimless and Rimmed)
6.5 x 48mm Sauer
7 x 57 mm Mauser
6 x 57mm Mauser
6.5 x 27mm
6.5 x 53.5mm Daudeteau

3. BRITISH SPORTING RIFLE CARTRIDGES

.255 Rook Rifle
.369 Purdey Nitro Express
.375 Rimless Wesley Richards Nitro
.400/360 Purdey Flanged
.450 Nitro Express
.450 No. 2 Nitro Express
.476 Nitro Express
.600 Nitro Express
.360 No. 2 Nitro Express
.375/.303 Axite
.400/.360 Wesley Richards
.450 Black Powder Express
.450 Nitro for Black Powder Express
.475 No. 2 Jeffrey
.500 Black Powder Express

4. PISTOL AND REVOLVER CARTRIDGES

2.7mm Kolibri Auto
4.25mm Liliput Auto
5mm Clement Auto
6.5mm Bergmann, standard rimless 7mm Baby Nambu
.30/7.65mm Borchardt
.38 Long Colt
.44 Smith & Wesson (Martin Primer)
3mm Kolibri Auto
5mm Bergmann
5.5 Velo Dog Short
6.5mm Bergmann (without extractor groove)7.65mm Roth-Sauer
.35 Smith & Wesson Auto
.44 Colt-Remington (Martin Primer)
.450 Long Revolver
.50 Remington (M71 Army)
9.8 Auto Colt
11mm German Service Revolver
11.75mm Montenegrin Revolver
.577 Pistol
11mm French Ordnance Revolver
11.25 Norwegian Raufoss (.45 ACP), pre-1946.

5. U.S. RIMFIRE CARTRIDGES

22 Rimfire cal. cartridges with experimental aluminum/steel cartridge cases.
.267 Rimfire cal. government experimental cartridge mfd. by Remington.
.32 Extra Short
.32 rimfire
.35 Allen
.38 Extra Long
.44 Short
.44 Long
.46 Long
.50 Peabody Musket
.50-70 Government
.54 Ballard
.56-50 Spencer
.56-52 Spencer, Tapered
.58 Joslyn Carbine
.58 Miller
.25 rimfire
.25 Short
.30 Long
.30 Shor.32 Long Rifle
.32 Extra Long
.38 rimfire
.41 Long
.44 Extra Long
.46 Extra Long
.46 Short
.50 Remington Navy
.52-70 Sharps
.56-46 Spencer
.56-52 Spener, Bottleneck
.56-56 Spencer
.58 Gatling Gun
.58 Mont Storm (Also known as .61 cal)

6. U.S. CENTERFIRE RIFLE CARTRIDGES

.22 Extra Long (Maynard)
.22-15-50 Stevens
.25-21 Stevens
.28 cal. centerfire experimental, mfd. at Frankford Arsenal.
.30-06 cal. cartridges mfd. by commercial manufacturers under government contract with headstamps indicating use for Palma Matches.
.30-30 Super X, Winchester "Bicentennial 76."
.30-30 Wesson
.308 Winchester, marked "Palma Match"
.32 Long Rifle
.32-30 Remington
.32-35 Maynard
.32-40 Winchester John Wayne .32-40 Bullard
.35 Winchester Self Loading

SECTION I

.38 Long
.38-35 Stevens
.38-50 Remington-Hepburn
.38-70 Winchester
.38-90 Winchester Express
.40-110 Winchester Express
.40-50 Sharps (Necked)
.405 Winchester
.40-60 Marlin
.40-60 Maynard 1882
.40-63 Ballard
.40-70 Sharps (Straight)
.40-70 Peabody "What Cheer"
.40-70 Remington
.40-72
.40-82
.40-90 Peabody "What Cheer"
.40-90 Sharps (Necked)
.40-90 Ballard
.44 Henry
.44 Extra Long Ballard
.44 Evans Long
.44-100 Wesson
.44-40 Extra Long
.44-60 Winchester
.44-70 Maynard
.44-77 Remington
.44-85 Wesson
.44-90 Remington (Stright 2.6" case)
.44-95 Peabody "What Cheer"
.45-50 Peabody (Sporting)
.40-65 Ballard Everlasting
.40-70 Sharps (Necked)
.22 Newton (Wildcat)
.25.36 Marlin
.25-25 Stevens
.28-30-120 Stevens
.30-03 Government (Note: Also known as Springfield)
.30-06 Commemorative, commemorating 50 years of the .30-06 cartridge as the standard U.S. service cartridge (headstamped USFA 1906-56).
.30-40 Wesson
.32 Ballard Extra Long
.32 Winchester Self Loading
.32-30 Maynard 1882
.32-35 Stevens
.32-40 Remington
.33 WCF
.35-40 Maynard
.38 Ballard Extra Long
.38-40 Remington-Hepburn
.38-45 Stevens
.38-50 Ballard
.38-56
.38-45 Bullard
.38-50 Maynard
.38-72
.40 Martin Primed
.40-40 Maynard 1882
.40-50 Sharps (Straight)
.70-150 Winchester
.40-70 Ballard
.40-70 Winchester
.40-75 Ballard
.40-85 Ballard
.40-90 Sharps (Straight)
.40-90 Ballard
.42 Martin Primer
.44 Wesson Extra Long
.44 Long C.F.
.44 Evans Short
.44-100 Ballard
.44-60 Sharps (Necked)
.44-60 Peabody "Creedmore"
.44-75 Ballard Everlasting
.44-77 Sharps
.44-90 Remington (Straight)
.44-90 Sharps (Necked)
.44-100 Remington "Creedmore" (Straight 2-6/10" case)
.45-75 Sharps (Straight) Note: Identical to .45-70 Govt. Cartridge
.45-70 Van Choate
.45-78-475 Wolcoff
.45-80 Sharpshooter
.45-100 Sharps (Straight)
.45-120 Sharps (Stright 3-1/4" case)
.50 cal. experimental machine gun cartridge with aluminum cartridge case
.50 U.S. Carbine
.50-119 Sharps
.50-140 Winchester Express
.50-50 Maynard
.50-90 Sharps
.55-100 Maynard
.58 U.S. Musket (Berdan)
.45-100 Ballard
.45-100 Remington
.45-125 Winchester
.50 Government
.50 Remington
.50-100 Sharps
.50-115 Bullard
.50-140 Sharps
.50-70 Muske
.50-95 Winchester
.58 Berdan Carbine
.60 cal. experimental cartridge
6mm Lee Navy

7. MISCELLANEOUS CARTRIDGES

- Duplex cartridges in all calibers with the exception of 7.62mm NATO.
- All cartridges intended for use in "squeeze bore" firearms; e.g., Gerlach taper bore ammunition.
- Dardick cartridges (Trounds) in all calibers.

SECTION II: Firearms Classified As Curios Or Relics Under 18 U.S.C. Chapter 44

The Bureau has determined that the following firearms are curios or relics as defined in 27 CFR 47 8.11 because they fall within one of the categories specified in the regulations.

Such determination merely classifies the firearms as curios or relics and thereby authorizes licensed collectors to acquire, hold, or dispose of them as curios or relics subject to the provisions of 18 U.S.C. Chapter 44 and the regulations in 27 CFR Part 478. They are still "firearms" as defined in 18 U.S.C. Chapter 44.

Albanian SKS semiautomatic rifles, caliber 7.62 x 39, manufactured in Albania from 1964 to 1978.
Alkartasuna, semiautomatic pistol, caliber .32.
All Original military bolt action and semiautomatic rifles mfd. between 1899 and 1946.
All properly marked and identified semiautomatic pistols and revolvers used by, or mfd. for, any military organization prior to 1946.
All shotguns, properly marked and identified as mfd. for any military organization prior to 1946 and in their original military configuration only.
Argentine D.G.F.M. (FMAP) System Colt Model 1927 pistols, marked "Ejercito Argentino" bearing S/Ns less than 24501.
Argentine D.G.F.M. - (F.M.A.P.) System Colt model 1927, cal. 11.25mm commercial variations.
Armand Gevage, semiautomatic pistols, .32ACP cal. as mfd. in Belgium prior to World War II.
Astra, M 800 Condor model, pistol, caliber 9mm parabellum
Astra, model 1921 (400) semiautomatic pistols having slides marked Esperanzo Y Unceta.
Astra, model 400 pistol, German Army Contract, caliber 9mm Bergmann-Bayard, S/N range 97351-98850.
Astra, model 400 semiautomatic pistol, cal. 9mm Bergmann-Bayard, second German Army Contract, in S/N range 92851 through 97350.
Auto-Mag pistols, calibers .44 AMP and .357 AMP, mfd. and/or assembled by Auto-Mag Corporation, TDE, OMC,
High Standard, Lee Jurras, or AMT from 1969 to 1985.
Auto Ordnance, West Hurley, NY, Korean War Commemorative Thompson semiautomatic rifle, caliber .45.
Auto Ordnance, West Hurley, NY, World War II Commemorative Thompson semiautomatic rifle, caliber .45.
Auto Ordnance Thompson, cal. 45 semiautomatic rifle, Vietnam Commemorative, S/Ns V0001-V1500, issued by the American Historical Foundation, Richmond VA.
Baker Gun and Forging Company, all firearms mfd. from 1899 to 1919.
Bannerman model 1937, Springfield rifle, caliber .30-06.
Bayard, model 1923 semiautomatic pistol, cal. 7.65mm or .380, Belgian manufacture.
Beretta, M 1951 pistol Israeli Contract, caliber 9mm parabellum.
Beretta, model 1915 pistols, cal. 6.35mm, 7.65mm, and 9mm Glisenti.
Beretta, model 1915/1919 (1922) pistol (concealed hammer), caliber 7.65mm.
Beretta, model 1919 pistol (without grip safety), caliber 6.35mm.
Beretta, model 1923 pistol, caliber 9mm Glisenti.
Beretta, model 1932 pistol, having smooth wooden grips w/"PB" medallion, cal. 9mm.
Beretta, model 1934 pistol, light weight model marked "Tipo Alleggerita" or "All" having transverse ribbed barrel, cal. 9mm.
Beretta, model 1934 pistols, cal. 9mm post war variations bearing Italian Air Force eagle markings.
Beretta, model 1934 pistols, cal. 9mm produced during 1945 or earlier and having S/Ns within the ranges of 500000-999999, F00001-F120000, G0001-G80000, 00001AA-10000AA, 00001BB-10000BB. The classification does not include any post war variations dated subsequent to 1945 or bearing postwar Italian proof marks.
Beretta, model 1935 pistol, Finnish Home Guard Contract, marked "SKY" on the slide, cal. 7.65mm.
Beretta, model 1935 pistol, Rumanian Contract, marked "P. Beretta - cal. 9 Scurt - Mo. 1934 – Brevet." on the slide, cal. 9mm.

SECTION II

Beretta, Model 92SB with detachable folding shoulder stocks, cal. 9mm S/Ns C31509Z through C31538Z.

FN Browning, model 1902 (usually known as the model 1903) semiautomatic pistol, caliber 9mm Browning long.

Browning .22 caliber, semiautomatic rifles, Grade II, mfd. by Fabrique Nationale in Belgium from 1956 to 1976.

Browning FN Medalist Gold Line and "Renaissance" engraved semiautomatic pistols, caliber .22.

Browning FN Model 1910 semiautomatic pistols, calibers .32 and .380, marked "Browning Arms Company."

Browning Superposed Bi-Centennial Ltd. Edition, shotgun.

Browning, "Baby" model pistol, Russian Contract, caliber 6.35mm.

Browning, Centennial model High Power Pistol, cal. 9mm parabellum.

Browning, Centennial model 92 lever action rifle, cal. .44 Magnum.

Browning, model 1906 Pocket Pistol if more than 50 years old.

Browning, model 1922 pistol, cal. 7.65mm or 9mm Kurz, marked "C.P.I.M." denoting issue to the Belgian Political Police.

Browning, model 1922 pistol, cal. 7.65mm, bearing German NSDAP or RFV markings.

Browning, M1935 Hi Power pistol, Canadian, Congolese, Indian and Nationalist Chinese Contracts, cal. 9mm parabellum.

Browning Hi Power, Classic Edition, 9mm caliber pistol, having S/Ns 245BC0001 through 45BC5000.

Browning High Power D-Day Commemorative, 9mm caliber pistol, having S/Ns 245DD0001 through 245DD00150.

Browning High Power, Gold Classic Edition, 9mm caliber pistol, having S/Ns 245GC0001 through 245GC0500.

Browning, Superposed Centennial, consisting of a 20 gauge superposed shotgun, supplied with an extra set of .30-06 cal. superposed barrels.

Budischowsky, model TP70, semiautomatic pistol, cal. .25 ACP, with custom S/N DB1.

Campo-Giro, model 1913 and 1913/16 pistol, cal. 9mm Largo.

Chinese Communist, types 51 and 54 (Tokarev) pistols, cal. 7.62mm.

Chinese, Peoples Republic of China, copy of German Walther PPK .32 ACP cal. w/Chinese proof marks, Type I and II.

Chinese, Peoples Republic of China, copy of Japanese Type Sigiura Shiki semiautomatic pistol, caliber 7.65mm.

Chylewski, semiautomatic pistol mfd. by S.I.G. Switzerland, cal. 6.35mm (.25 ACP).

Clement, pistol, Belgian manufacture, cal. 5mm Clement.

Colt, Model 1911A1, cal. .45, semiautomatic pistols, Famous Generals of WWII, S/Ns WWIIG001 – WWIIG200.

Colt .38 National Match semiautomatic pistol, all S/Ns and in their original configuration.

Colt .38 Special Kit semiautomatic pistols, mfd. from 1964-1970, S/Ns 00100H-00434H.

Colt .45 ACP Kit semiautomatic pistols, mfd. from 1964-1970, S/Ns 001000-011640.

Colt, factory engraved for "Dr. Ramon Grau San Martin," President of Cuba, .45 cal. pistol with 5" barrel and blue steel finish; S/N C231769.

Colt, "Duke," Commemorative, .22 caliber revolver.

Colt, Ace semiautomatic pistol, cal. .22, mfd. by Colt from 1931 to 1947, S/N range from 1 to 10935 including those marked "UNITED STATES PROPERTY" on the right side of the frame.

Colt, Ace Service model semiautomatic pistol, cal. .22, mfd. by Colt from 1935 to 1945, S/N range from SM1 to SM13803 including those marked "UNITED STATES PROPERTY" on the right side of the frame.

Colt, Age of Flight 75th Anniversary semiautomatic pistols, cal. .45.

Colt, Aircrewman revolver produced between 1951 and 1959, cal. .38 Special, marked "Property of U.S. Air Force" on back strap, having Air Force issue numbers of 1 - 1189 and in the S/N range 1902LW – 90470LW.

Colt, Alabama Sesquicentennial, .22.

Colt, Alamo, .22 and .45.

Colt, American Combat Companion Officers model, cal. .45ACP pistol marked "1911 American Combat Companion 1981, 70 Years at America's Side."

SECTION II

Colt, American Combat Companion, Enlisted Man's model, cal. .45ACP pistol marked "1911 American Combat Companion 1981, 70 Years at America's Side."

Colt, American Combat Companion, General Officers model, cal. .45ACP pistol marked "1911 American Combat Companion 1981, 70 Years at America's Side." S/Ns 1 STAR, 2 STAR 3 STAR, 4 STAR, and 5 STAR.

Colt, Appomattox Court House Centennial, .22 and .45.

Colt AR-15 Sporter "The Viet Nam Tribute Colt Special Edition" .223 cal. semiautomatic rifle, bearing the American Historical Foundation registry numbers of VT0001 through VT1500.

Colt, Argentine model 1927, pistols, cal. .45, commercial variations.

Colt, Arizona Ranger model commemorative, .22 Revolver.

Colt, Arizona Territorial Centennial, .22 and .45.

Colt, Arkansas Territory Sesquicentennial, .22.

Colt, Army Model double action revolver, any cal., mfd. between 1899 and 1907.

Colt Army Special Model revolver, .32-20 cal., having a barrel length of 4'1/2", with factory engraving, S/N 329653.

Colt, ATF Special Edition, Deluxe model automatic pistol, cal. .45 ACP.

Colt, ATF Special Edition, Python Revolver, caliber .357 magnum.

Colt, ATF Special Edition, Standard model automatic pistol, cal. .45 ACP.

Colt, Abilene, .22 (Kansas City-Cow Town).

Colt, Bat Masterson, .22 and .45 (Lawman Series).

Colt, Battle of Gettysburg Centennial, .22.

Colt, Belleau Wood, .45 Pistol, (World War I Series).

Colt, Border Patrol model Revolver, .38 Special Heavy Duty, Police Positive (D) style frame, S/Ns within the range 610000 through 620000.

Colt, Buffalo Bill Historical Center, Winchester Museum, Special Issue, Colt Single Action revolver, cal. .44-40, S/N 21BB.

Colt, Buffalo Bill Wild West Show single action army .45 caliber.

Colt, California Bicentennial, .22.

Colt, California Gold Rush, .22 and .45.

Colt, Camp Perry Single Shot Target Pistols, .22 long rifle or .38 Special caliber.

Colt, Carolina Charter Tercentenary, .22 and .22/.45.

Colt, Chamizal Treaty, .22 and .45.

Colt, Chateau Thierry, .45 Pistol, (World War I Series).

Colt, Cherry's Sporting Goods 35th Anniversary, .22/.45.

Colt, Chisholm Trail, .22 (Kansas Series-Trails).

Colt, Civil War Centennial Single Shot, .22.

Colt, Coffeyville, .22 (Kansas Series-Cow Town).

Colt, Colorado Gold Rush, Colt, Colonel Samuel Colt, Sesquicentennial, .45.

Colt, Colt's 125th Anniversary, .45.

Colt, Columbus (Ohio) Sesquicentennial, .22.

Colt, Custom Gun Shop's "Custom Edition Sheriff's model" Single Action Revolver, cal. .45 Colt, S/Ns 1 to 35.

Colt, DA .38, New Army and Navy Revolver, made from 1899 to 1907.

Colt, Dakota Territory, .22.

Colt, Des Monies, Reconstruction of Old Fort, .22 and .45.

Colt, Detective Special revolver, cal. .38, S/N 418162, owned by Colonel Charles A. Lindbergh.

Colt, Detective Special revolvers, 2" barrels, marked "R.M.S. P.O. DEPT" or "U.S.P.O. DEPT."; S/Ns above 467000.

Colt, Dodge City, .22 (Kansas Series-Cow Town).

Colt, European Theater, .45 Pistol (World War II Series).

Colt, First model, Match Target Woodsman, cal. .22, semiautomatic pistol, mfd. from 1938 to 1944, S/Ns MT1 to MT15,000.

Colt, Florida Territory Sesquicentennial, .22.

Colt, Fort Findlay (Ohio) sesquicentennial, .22.

Colt, Fort Hays, .22 (Kansas Series-Forts).

Colt, Fort Larned, .22 (Kansas Series-Forts).

Colt, Fort McPherson (Nebraska) Centennial Derringer, .22.

SECTION II

Colt, Fort Scott, .22 (Kansas Series-Forts).
Colt, Fort Stephenson (Ohio) sesquicentennial, .22.
Colt, Fourth model Derringer, cal. .22 short rimfire, cased as a set of two pistols in a leather book titled "Colt Derringer, Limited Edition Colt, General George Meade, Pennsylvania, by Colt," on the spine of the book and "A Limited Edition by Colt," on the cover.
Colt, Forty-Niner Miner, .22.
Colt, General George Meade, Pennsylvania Campaign, .22 and .45.
Colt, General Hood, Tennessee Campaign Centennial, .22.
Colt, General John Hunt Morgan, Indiana Raid, .22.
Colt, General Nathan Bedford Forrest, .22.
Colt, Geneseo (Illinois) 125th Anniversary, Derringer, .22.
Colt, Golden Spike Centennial, .22.
Colt, Government model pistols in cal. .45 ACP, BB series.
Colt, H. Cook, "1 to 100," .22/.45.
Colt, Idaho Territorial Centennial, .22.
Colt, Indiana Sesquicentennial, .22.
Colt, J frame, Officers model Match, .38 Special revolver mfd. from 1970-1972, identified by J S/N prefix.
Colt, Joaquin Murrieta, "1 of 100," .22/.45.
Colt, John M. Browning Commemorative, .45 cal., semiautomatic pistol, S/Ns JMB 0001 – JMB 3000, and numbers GAS O JMB, PE CEW JMB, and 0003JMB.
Colt, John Wayne, Commemorative, .45 long Colt caliber revolver.
Colt, Kansas Centennial, .22.
Colt, Lightning model double action revolver, any cal. mfd. between 1899 and 1909.
Colt, Lightning rifles mfd. in 1899 through 1904.
Colt, Lord and Lady Derringer, .22 cal., as mfd. by Colts Patent Firearms Manufacturing Co., Hartford, CT.
Colt, Los Angeles Police Department (L.A.P.D.) Special Edition .45 cal. Government model semiautomatic pistol.
Colt, Maine Sesquicentennial, .22 and .45.
Colt, Mark IV, Government Model, commemorative "Michigan State Police 60th Anniversary, 1917-1977." The left side of slide engraved with scroll pattern and depicts 4 modes of transportation; Horse, Motorcycle, Auto and Helicopter S/Ns 1 to 1608.
Colt, Match Target Woodsman Semiautomatic Pistol, cal. .22LR., S/N 128866S, owned by Ernest Hemingway.
Colt, Meuse Argonne, .45 Pistol, (World War I Series).
Colt, Missouri Sesquicentennial single action army .45 caliber.
Colt, Missouri Sesquicentennial, .22.
Colt, Mk IV Series 70 semiautomatic pistols in all cals., which were incorrectly marked at the factory with both Colt Government model markings and Colt Commander markings.
Colt, model 1873 Peacemaker Centennial 1973, single action revolver, .44/.40 or .45.
Colt, model 1900 semiautomatic pistol, cal. .38, in original configuration.
Colt, model 1902 semiautomatic pistol, military model, cal. .38, in original configuration.
Colt, model 1902 semiautomatic pistol, sporting model, cal. .38, in original configuration.
Colt, model 1903 Pocket (exposed hammer), semiautomatic pistol cal. .38 ACP.
Colt, model 1903 Pocket (hammerless), semiautomatic pistol, cal. .32.
Colt, model 1908 Pocket (hammerless), semiautomatic pistol, cal. .380.
Colt, model 1908, cal. .25 ACP, hammerless semiautomatic pistol, having a grip safety, in S/N range 1 - 409061.
Colt, model 1911, commercial semiautomatic pistols, cal. .45 ACP, S/Ns Cl - C130000.
Colt, model 1911-A, commercial model, in cal. .45 and bearing Egyptian inscription meaning police, on the upper forward right-hand side of the trigger guard and having S/Ns within the range of C186000 to C188000.
Colt Model "Courier" double action revolver, .32 caliber, with 3-inch barrel.
Colt Model "Marshal" double action revolver, .38 special caliber, 2-inch and 4-inch barrels.
Colt, 50th Anniversary of the Battle of the Bulge, Model 1911A1, .45 caliber, semiautomatic pistol, having S/Ns between BB001 and BB300.

SECTION II

Colt, Montana Territory Centennial, .22 and .45.
Colt, National Match semiautomatic pistols, all S/Ns, in original configuration.
Colt, Nebraska Centennial, .22.
Colt, Ned Buntline Commemorative, cal. .45 revolver.
Colt, Nevada Centennial "Battle Born," .22 and .45.
Colt, Nevada Centennial, .22. and .45.
Colt, New Frontier .22 LR Revolvers, "Kit Carson" Commemorative, Colt model GB275.
Colt, New Frontier and Single Action Army model revolvers originally ordered & shipped with factory engraving, accompanied by a letter from the manufacturer confirming the authenticity of the engraving.
Colt, New Frontier, .357 magnum cal., single action revolver, barrel length 4-3/4" S/N 4411NF.
Colt, New Frontier, .45 cal., Abercrombie and Fitch, "Trailblazer."
Colt, New Jersey Tercentenary, .22 and .45.
Colt, New Mexico Golden Anniversary, .22.
Colt, New Police revolvers, .32 Colt cal., S/Ns 1 through 49,50.
Colt, New Service revolvers as mfd. between 1899 - 1944, all variations, all calibers.
Colt, NRA Centennial, Gold Cup National Match pistol, in cal. .45.
Colt, NRA Centennial, single action revolver, in cals. .357 Magnum and .45.
Colt, Officers model (1904-1930), .38 cal. revolver.
Colt, Officers model (1930-1949), .22 cal. revolver.
Colt, Officers model Match (1953-1969), .22 and .38 cal. revolvers.
Colt, Officers model Special (1949-1952), .22 and .38 cal. revolvers.
Colt, Officers model Target (1930-1949), .32 and .38 cal revolvers.
Colt, Officer's Model, .38 caliber revolver, S/Ns535472, 585683.
Colt Official Police Model revolver, cal. .32-20, having a barrel length of 5 inches, with factory engraving by Wilbur Glahn, S/N 554399.
Colt Official Police Model revolver, .38 Colt Special cal., having a barrel length of 6 inches, with factory engraving by Wilbur Glahn, S/N 554445.
Colt, Official Police Revolver, cal. .38, Silver Inlaid and Engraved by Wilbur A. Glahn, S/N 583469.
Colt, Oklahoma Territory Diamond Jubilee, .22.
Colt, Oregon Trail, .22 (Kansas Series-Trails).
Colt, Pacific Theater, .45 Pistol (World War II Series).
Colt, Pat Garrett, .22 and .45 (Lawman Series).
Colt, Pawnee Trail, .22 (Kansas Series-Trails).
Colt, Peacemaker Commemorative, .22 and .45 revolver.
Colt, Pocket Positive revolver, .32 cal.
Colt, Pocket Positive revolver, S/N 6164.
Colt, Police Positive .38 cal. revolvers, 2" barrels, marked "R.M.S. P.O. DEPT." or "U.S.P.O. DEPT."; S/Ns above 383000.
Colt, Police Positive revolver (1909); "St. L.P.D. #466" on butt; S/N 24466.
Colt, Pony Express Centennial, .22.
Colt, Pony Express, Russell, Majors and Waddell, Presentation model .45.
Colt, Python Model revolver, .357 cal., having a barrel length of 3 inches, with factory Type A engraving by Denise Thirion, in a presentation case with accessories, S/N T21895.
Colt, Python Revolver, cal. .357 Magnum, engraved and inlaid with the Crest of the United Arab Emirates.
Colt, Revolver, cal. .38, Police Positive, S/N 139212.
Colt, Rock Island Arsenal Centennial Single Shot, .22.
Colt, Santa Fe Trail, .22 (Kansas Series-Trails).
Colt model San Jacinto Special edition single action army revolver, .45 caliber, marked "THE BATTLE OF SAN JACINTO, APRIL 21, 1836" and "REMEMBER GOLIAD," 200 produced.
Colt, Second (2nd) Marne, .45 Pistol (World War I Series).
Colt, Shawnee Trail, .22 (Kansas Series-Trails).
Colt, Sheriffs model revolver, cal. .44 and .45.
Colt, Single Action Army (Bisley, Standard, and target variations), all original, mfd. from 1899 to 1946, S/N range from 182000 to 357869.

Colt, Single Action Army revolver, cal. .45, S/N 85163A, engraved and inlaid with a bust of President Abraham Lincoln.
Colt , Single Action Army (2nd Generation) revolvers, having S/Ns from 0001SA to 82000SA, all calibers, made between 1956 and 1976.
Colt, Single Action Revolvers, cal. .45, engraved & silver inlaid for presentation to Chuck Connors, S/Ns CC1 and CC2.
Colt, St. Augustine Quadricentennial, .22.
Colt, St. Louis Bicentennial, .22 and .45.
Colt Government model Texas Battleship Special Edition, .45 caliber pistols marked "BATTLESHIP EDITION - USS TEXAS," 500 produced.
Colt, Texas Ranger, .45.
Colt, Texas Sesquicentennial Standard and Premier model single action army, .45 cal.
Colt, the Liege Number 1 Colt Single Action Army Revolver, cal. .45, S/N Liege No. 1.
Colt, "The Right to Keep and Bear Arms" commemorative, .22 cal. Peacemaker Buntline, single action revolver having a 7-1/2" barrel with the inscription, "The Right to Keep and Bear Arms" inscribed on the barrel and a S/N range of G0001RB - G3000RB.
Colt, Theodore Roosevelt single action army .44-40 cal.
Colt, United States Bicentennial Commemorative, Python revolver, caliber .357.
Colt, United States Bicentennial Commemorative single action army revolver cal. .45.
Colt, West Virginia Centennial, .22 and .45.
Colt, Wichita, .22 (Kansas Series-Cow Town).
Colt, Wild Bill Hickok, .22 and .45 (Lawman Series).
Colt model Wisconsin State Patrol, 45th Anniversary, Special Edition Commemorative, 70 series, .45 government model pistols, consecutively numbered 1 through 126.
Colt, Woodman, cal. .22, semiautomatic target pistol, mfd. from 1915 to 1943, S/Ns 1 to 157000.
Colt, Woodsman, .22 caliber semiautomatic pistols, all models, all series, all S/Ns, (to include Match Target, Challenger, Huntsmen, and Targetsman), made prior to 1978.
Colt, Woodsman, First Model Match Target, .22 cal. semiautomatic pistols mfd. in or before 1944, a and having S/Ns MT1 through MT15100.
Colt, Wyatt Earp, .22 and .45 (Lawman Series).
Colt, Wyatt Earp, Buntline Special, .45 (Lawman Series).
Colt, Wyoming Diamond Jubilee, .22
Colt XIT Special Edition Single Action Army revolver, .45 caliber, barrel chemically etched with a covered wagon scene and "XIT RANCH," and marked "1 of 500."
Colt, 150th Anniversary single action army buntline, .45 cal.
Czechoslovakian, CZ27, 7.65mm semiautomatic pistol with Nazi markings.
Coston Supply Co., Coston Line Throwing Gun, .45-70 cal. S/N 447.
Czechoslovakian, CZ38, pistol cal. .380ACP.
Czechoslovakian, CZ50 pistol, cal. 7.65mm.
Czechoslovakian, CZ52 pistol, cal. 7.62mm.
Czechoslovakian, model 1952 and 1952/57, 7.62 x 45mm and 7.62 x 39mm cal., semiautomatic rifles.
Czechoslovakian, VZ-54 sniper rifle, manufactured from 1954 through 1957, caliber 7.62 x 54R, all S/Ns with scope mount and 2.5 power telescope.
Czechoslovakian, VZ82/CZ82 all calibers, all serial numbers.
* Daisy, Model V/L, .22 caliber caseless rifle, manufactured during 1968-1969.
Danish, M1910/1921 Bayard, pistol, cal. 9mm Bergmann-Bayard.
Davis Warner, Infallible, semiautomatic pistol, cal. .32.
Dreyse, semiautomatic pistols, all calibers.
Egyptian Raschid, semiautomatic rifle, cal. 7.62 x 39mm, original Egyptian military production.
Egyptian, Hakim (Ljungman) 7.92mm semiautomatic rifle as mfd. in Egypt.
Erma-Werke, Model EL 24, cal. .22, rifle, mfd. prior to 1946.
Esser-Barraft, English manufacture, slide action rifle, cal. .303.
Fabrique Nationale 1889-1989 Centenary High Power pistols, cal. 9mm.
Fabrique Nationale, Model 1906, .25 caliber semiautomatic pistol, all S/Ns.
Fabrique Nationale, model SAFN49 semiautomatic rifles, any caliber.
FN F.A.L. G and GL series, semiautomatic rifles, imported by Browning Arms Company, Arnold,

MO from 1959 to 1963, with the following S/Ns: G Series: G492, G493, G494, G537-G540, G649-G657, G662-G673, G677-G693, G709-G748, G752-G816, G848-G1017, G1021, G1033, G1035, G1041, G1042, G1174-G1293, G1415-G1524, G1570-Gl784, G1800-Gl979, G1981-G1995, G2247-G2996, G3035-G3134. GL Series: GL749, GL835, GL1095-GL1098, GL1163,GL1164, GL1165, GL2004-GL2009, GL3135-GL3140.

Firepoint International, Ltd. Of England, 7.92x33 cal. S/Ns 08, 10, 11, 12, 14, 15, 16.

French Military Rifle Model 1949/56, in 7.62 x 51mm (NATO) cal. French, model 1949, cal. 7.5mm, semiauto. rifle (Fusil Mle. 1949 (MAS) 7.5mm).

French, model 1949/56 (Fusil Mle (MAS 7.5mm)) semiautomatic rifle.

A.H. Fox, Double barrel shotguns, all gauges, all grades, mfd. By Ansley H. Fox, Philadelphia, PA, and Savage Arms, Utica, NY, from approx. 1907 - 1947.

Geha and Rmo, shotguns made from Mauser rifles after World War I prior to 1946.

German military training rifles, cal. .22, single shot and repeaters, all manufacturers, in their original military configuration, marked "Kleinkaliber Wehersportsgewehr" (KKW), mfd. prior to 1946.

German sporting rifles, cal. .22, sporting rifles, single shot and repeaters, all manufacturers, in original configuration, marked "Deutsche Sportmodell" (DSM), prior to 1946.

German, model 1916 Grenatenwerfer original spigot type mortars.

German, P38 pistols, cal. 9mm parabellum mfd. prior to 1947.

Greener, Martini action, 14 gauge shotgun.

Gustloff, semiautomatic pistol in cal. 7.65mm mfd. by Gustloff Werke, Suhl, Germany.

Hammond or Grant Hammond, pistols, all models, variations or prototypes, made by Grant Hammond Corporation, New Haven, CT.

Hammond/Hi-Standard, semiautomatic pistols, in caliber .45.

Harrington and Richardson "Reising" Model 60 semiautomatic rifles, .45 ACP caliber, manufactured between 1944 and 1946.

Harrington and Richardson (H&R), Abilene Anniversary, .22 revolver.

Harrington and Richardson (H&R) Handy Gun, pistols with original rifled barrel, mfd. At Worcester, MA, all calibers, all barrel lengths, without shoulder stock.

Harrington and Richardson (H&R) Handy guns, mfd. at Worcester, MA, with shoulder stock, having an original smoothbore barrel 18 inches in length or greater, or original rifled barrel 16 inches in length or greater.

Harrington & Richardson Trapdoor Springfield carbine, .45-70 caliber, 100th Anniversary Little Big Horn Commemorative, manufactured between 1973-1981.

H&R, Centennial Officer's model Springfield rifle .45-70 Government.

H&R, Centennial Standard model Springfield rifle .45-70 Government.

H&R, model 999, revolver, cal. 22 Long rifle, barrel 6", 110th year commemorative, S/Ns from 001 to 999.

H&R, self loading semiautomatic pistol, caliber .32.

Hartford Arms and Equipment Company, single shot target pistol, caliber .22LR.

Hartford Arms and Equipment Company, repeating pistol, cal. .22LR.

Hartford Arms and Equipment Company, model 1928 pistol, cal. .22LR.

Hi-Standard, experimental electric free pistol, cal. .22 long rifle.

Hi-Standard, experimental electric free pistol, cal. .38 special.

Hi-Standard, experimental electric free pistol, cal. .38 special.

Hi-Standard, model P 38, semiautomatic pistol, cal. .38 SPL.

Hi-Standard, experimental model T-3 semiautomatic pistol, cal. 9mm Luger.

Hi-Standard, experimental ISU rapid fire semiautomatic pistol, caliber .22 short.

High Standard, Second Model Olympic pistol, cal. .22 short, mfd. between 1951 - 1953, S/Ns 330,000 – 439,999.

High Standard, Crusader Commemorative, Deluxe Pair, .44 Magnum and .45 Colt revolvers.

High Standard, model A pistol, caliber .22LR.

High Standard, model B pistol, caliber .22LR.

High Standard, model C pistol, caliber .22 Short.

High Standard, model D pistol, caliber .22LR.

High Standard, model E pistol, caliber .22LR.

High Standard, model H-A pistol, caliber .22LR.

High Standard, model H-B pistol, first model, caliber .22LR.
High Standard, model H-B pistol, second model, caliber .22LR.
High Standard, model H-D pistol, caliber .22LR.
High Standard, model H-E pistol, caliber .22LR.
High Standard, model USA-HD pistol, caliber .22LR.
High Standard, model HD-Military pistol, caliber .22LR.
High Standard, model G-380 pistol, caliber .380.
High Standard, model G-B pistol, caliber .22LR.
High Standard, model G-D pistol, caliber .22LR.
High Standard, model G-E pistol, caliber .22LR.
High Standard, model G-O (First model Olympic) pistol, caliber .22 Short.
High Standard, Supermatic Trophy model 107, .22 pistol Olympic Commemorative model.
High Standard, 1980 Olympic Commemorative, .22 caliber semiautomatic pistols, S/Ns USA1 - USA1000.
Holland and Holland, Royal Double Barrel Shotgun, .410 Gauge, S/N 36789.
Hopkins and Allen, model 1901, "FOREHAND," caliber .32 S&W long.
Hopkins and Allen, Revolver, .32 cal., S/N G 9545.
Hungarian Model 48 (Mosin Nagant M44 type) carbine, caliber 7.62 x 54R, manufactured in Hungary, and identified by the manufacturer code 02 on the chamber area and marked with the date of manufacture in the 1950's or earlier.
India rifle, cal. 7.62 2A, all variations, originally mfd. At Ishapore Arsenal, India prior to 1965.
Italian, Brixia, M1906, pistol, cal. 9mm Glisenti.
Italian, Glisenti, M1910, pistol, cal. 9mm Glisenti.
* Ithaca, Bicentennial Model 37, pump action shotguns, S/Ns USA0001 to USA1976.
Ithaca, double barrel shotguns actually mfd. in NY by the Ithaca Gun Co. Ithaca, NY. All gauges and all models, having barrels at least 18" in length and an overall length of at least 26", mfd. before 1950.
Ithaca Gun Co., single barrel trap guns, break open all gauges, all models actually mfd. at Ithaca, NY, before 1950.
Ithaca, St. Louis Bicentennial, model 49, .22 rifle.
Iver Johnson Arms, Pistol, cal. .380, U.S. Border Patrol 60th Anniversary commemorative, S/Ns USBP 0001 to USBP 5000.
Iver Johnson Arms, model M 1 Carbine, cal. .30, Korean War commemorative, S/Ns KW0001 to KW2500.
Iver Johnson Arms, model M 1 Carbine, cal. .30, Airborne commemorative, S/Ns KW0001 – KW2500.
Jieffeco, pistol, Belgian manufacture, caliber 7.65mm.
Jieffeco, semiautomatic pistol, cal. .25 ACP, marked "Davis Warner Arms Corp., N.Y."
Kimball, pistols, all models, all calibers.
Kolibri, pistols, cals. 2.7mm and 3mm Kolibri.
L. C. Smith, Shotguns mfd. by Hunter Arms Co. and Marlin Firearms Co. from 1899 to 1971.
Langenhan, semiautomatic pistols, all calibers.
Lahti, L-35 pistol, Finnish manufacture, caliber 9mm parabellum.
Lahti Swedish Model M40 pistols, cal. 9mm, all variations, mfd. prior to 1968.
Lee Enfield, No. 1 Mk III bolt action rifle, cal. .303, mfd. in Ishapore, India between 1946 and 1960. Original military configuration only.
Lee Enfield Rifle, caliber 7.62 2A and 2A1 "India" all variations, originally manufactured at Ishapore Arsenal, India, through 1973.
Lefever, shotguns made from 1899 to 1942.
Luger, Model 1902 Cartridge Counter, Mauser commercial, semiautomatic pistol, cal. 9mm, mfd. 1982.
Luger, pistol, all models and variations mfd. prior to 1946.
Luger, Mauser commercial manufacture, semiautomatic pistol, 70 Jahre, Parabellum-Pistole, Kelsoreich Russiand, commemorative, cal. 9mm.
Luger, Mauser commercial manufacture, semiautomatic pistol, 76 Jahre, Parabellum-Pistole, 1900-1975, commemorative, cal. 7.65mm.
Luger, Mauser commercial manufacture, semiautomatic pistol, 75 Jahre, Parabellum-Pistole,

Konigreich Bulgarian, commemorative, caliber 7.65mm.
Luger, Mauser Parabellum, semiautomatic pistol, 7.65mm or 9mm Luger, 4 and 6" barrel, Swiss pattern with grip safety and the American Eagle stamped on the receiver; made from 1970 to 1978.
MAB, model R pistol, caliber 9mm parabellum.
Makarov, pistol, Russian and East German, caliber 9mm Makarov.
Mannlicher, pistol, M1900, M1901, M1903 and M1905, caliber 7.63mm Mannlicher.
Marlin, Model 336 TS carbine, cal. 30-30, Powell Wyoming 75th anniversary commemorative, having PW S/N prefix.
Marlin, 90th Anniversary, model 39-A, .22 rifle.
Marlin, 90th Anniversary, model 39-A, .22 carbine.
Mauser, semiautomatic pistols mfd. prior to 1946, any caliber.
Mauser, Congolese model 1950 rifles marked FP 1952 on the receiver, caliber .30/06.
Mauser Luger, S/N 11.010034, 9mm, special engraving and ivory grips.
Mauser Luger, S/N RG 900/1001, 9mm, special engraving and walnut grips.
Mauser, model 1935 rifle 7 x 57mm cal. with Chilean Police Markings.
Mauser, P 38, pistols caliber 9mm, marked SVW46.
Mauser, rifles, bolt action and semiautomatic any caliber, commercially produced by Waffenfabrik Mauser, Oberndoff, Germany, prior to 1945.
MBA Gyrojet Carbine, S/N B5057.
MBA Gyrojet semiautomatic pistols, cal. 12mm or smaller, all models.
Menz, Liliput, German manufacture caliber 4.25mm.
Menz, PB III, in cal. 7.65mm, mfd. by August Menz, Suhl, Germany.
Menz, PB IIIA, in cal. 7.65mm, mfd. by August Menz, Suhl, Germany.
Menz, PB IV, in cal. 7.65mm, mfd. by August Menz, Suhl, Germany.
Menz, PB IVa, in cal. 7.65mm, mfd. by August Menz, Suhl, Germany.
Menz, Special, in cal. 7.65mm, mfd. by August Menz, Suhl, Germany.
Mexican, Obregon, pistol, caliber .45 ACP.
Mexican Model 1954 bolt action Mauser rifles and carbines, caliber .30-06 (original military configuration only).
Mossberg, Model 40, cal. .22 tubular fed, bolt action rifles mfd. from 1933-1935.
Mossberg, Model 42TR Targo, cal. .22 smoothbore, bolt action rifles, mfd. from 1940-1942.
Mossberg, Model 479RR "Roy Rogers" limited edition, cal. .30-30 lever action rifles, total number mfd. in 1983 only.
Mossberg, Model 25, .22 caliber bolt action rifles.
Mugica, model 120, pistol, caliber 9mm parabellum.
Navy Arms, Oklahoma Diamond Jubilee Commemorative, Yellow Boy Carbine.
Norinco (Chinese) AK47S, 5.56x45mm caliber, S/N 403876.
Norinco (Chinese) AK47S84S-1, 5.5 6x 45mm caliber, S/N 303052.
Norinco (Chinese) AK47S, 7.62 x 39mm caliber, S/N 1620127.
North Korean Type 1964, pistol, caliber 7.62mm Tokarev.
North Korean, Type 68, 7.62 x 25mm caliber semiautomatic pistols.
O.F. Mossberg Model 472SBAS "American Indian Commemorative" .30-30 and .35 Remington caliber lever action rifles, having Indian scenes etched on receiver, manufactured in 1974 only.
Ortgies, semiautomatic, caliber .25, with S/N 10073.
Ortgies, semiautomatic, caliber .32, with S/N 126314. OWA, semiautomatic pistol, caliber .25.
PAF, "Junior" semiautomatic pistol, caliber .25, mfd. by the Pretoria Arms Factory Ltd. of South Africa.
PAF, pistol, marked "BRF," caliber .25, mfd. by the Pretoria Arms Factory Ltd. of South Africa.
Parker-Hale, Model T-4, cal. 7.62mm, bolt action target rifles, mfd. prior to 1975.
Parker, shotguns, all grades, all gauges, produced by Parker Brothers, Meridan, CT, and Remington Arms, Ilion, NY, from 1899 through 1945.
Pedersoli, 120th Anniversary of the Remington Creedmore, .45/70 caliber, single shot rolling block rifle, having S/Ns between CR001 and CR300.
Pedersoli, 125th Anniversary of the Springfield, Model 1873 Trapdoor rifle, .45/70 caliber, having S/Ns between 125th- 001 through 125th-125.

SECTION II

Pedersoli, Sharps Creedmoor, single shot rifles, caliber .45-70, S/Ns SCR001 through SCR300.

* Pedersoli, Springfield Officer's Model trapdoor rifle, .45-70 caliber, having S/Ns 2NFM001 through 2NFM250.

Phoenix, (U.S.A.), pistol, caliber .25 ACP.Polish Mosin Nagant M44 type carbines, caliber 7.62 x 54R, manufactured in Poland, identified by the manufacturer code 11" in an oval on the chamber area, and marked with the actual date of manufacture during the 1950s or earlier.

Poly Tech (Chinese) AK 47S (386), 7.62 x 39mm caliber, S/N P47-11545.

James Purdey, Over & Under shotgun, 12 gauge, S/N 26819, engraved and gold inlaid.

Reising, .22 caliber, semiautomatic pistol.

Remington, No. 1, Mid Range Rolling Block target rifle reproduction, caliber .45-70, S/Ns beginnning with "RB97."

Remington, Model 31, pump action shotguns, 12, 16, and 20 gauge, mfd. 1931 - 1950.

Remington Rolling Block firearms, all models mfd. from 1899 - 1935.

Remington, over/under Derringer, caliber .41 rim fire, Remington Arms Company, Ilion, NY, made between 1898 and 1935.

Remington, Canadian Territorial Centennial, model 742, rifle.

Remington, model 12, rifle, cal. .22 short, long rifle, and .22 Remington Special mfd. by Remington Arms, Union Metallic Cartridge Co., Remington Works, Ilion, NY, from 1909 to 1936.

Remington, model 30 rifles.

Remington, model 720 rifles.

Remington, model 51, semiautomatic pistol, cals. .32 ACP or .380 ACP.

Remington, Montana Territorial Centennial, model 600, rifle.

Remington, 150th Anniversary model Nylon 66 semiautomatic rifle, caliber .22LR.

Remington, 150th Anniversary model 1100SA semiautomatic shotgun, caliber 12 gauge.

Remington, 150th Anniversary model 552A semiautomatic rifle, caliber. 22LR.

Remington, 150th Anniversary model 572A slide action rifle, caliber .22LR.

Remington, 150th Anniversary model 742ADL semiautomatic rifle caliber .30/06.

Remington, 150th Anniversary model 760ADL slide action rifle caliber .30/06.

Remington, 150th Anniversary model 870SA slide action shotgun, caliber 12 gauge.

Rheinmetal, semiautomatic pistols, caliber .32.

Rhode Island Arms Co., Morrone Model 46, shotgun.

* Romanian Model 56 semiautomatic carbine (SKS), caliber 7.62 x 39mm in original configuration as manufactured at Uzina Mechanica Cugir, Romania, from 1956 through 1962, having original Romanian S/Ns consisting of two letters followed by up to four digits. The right side of the receiver is also marked with the year of production and the manufacturer's trademark consisting of an unfletched arrow within a triangle.

Roth Steyr, 1907, semiautomatic pistol, caliber 8mm.

Romanian AK 47S, 5.45 x 45mm caliber , S/N 3-040053-97.

Ruger, Blackhawk .44 with 6-1/2" barrel, revolver with S/Ns 1 to 29860.

Ruger, Blackhawk .44 with 7-1/2" barrel, revolver with S/Ns 17000 to 29860.

Ruger, Blackhawk .44 with 10" barrel, revolver with S/Ns 18000 to 29860.

Ruger, Blackhawk .357 with 4-5/8" barrel, revolver with S/Ns 1 to 42689.

Ruger, Blackhawk .357 with 6-1/2" barrel, revolver with S/Ns 20000 to 42689.

Ruger, Blackhawk .357 and 9mm stainless steel revolver with S/Ns 32-56000 - 32-59000.

Ruger, Blackhawk .357 magnum with 10" barrel, revolver with S/Ns 20000 - 38000.

Ruger, Canadian Centennial, Matched No. 1 Rifle Sets, Special Deluxe.

Ruger, Canadian Centennial, Matched No. 2 Rifle Sets.

Ruger, Canadian Centennial, Matched No. 3 Rifle Sets.

Ruger, Canadian Centennial, model 10/22, carbine.

Ruger, l0/22 Canadian Centennial, carbine with S/Ns Cl to C4500.

Ruger, Falling Block Long Range Creedmore rifle, cal. .45 (Sharps), S/N 130-06888, The Amber Silver Jubilee.

Ruger, flattop, "Blackhawk" revolvers, cals. .44 Magnum and .357 magnum, all barrel lengths, made from 1955 through 1962.

Ruger, flattop, single-six, .22 cal. revolvers with flat side loading gate, all barrel lengths, made from 1953 through 1956.

SECTION II

Ruger, Hawkeye, pistol with S/Ns 1 to 3296.
Ruger, Lightweight Single Six, Revolver with S/Ns 200000 to 212630.
Ruger, Mark I "U.S." stamped medallion, pistol with S/Ns 76000 - 79000.
Ruger, Single Six, engraved revolver with S/Ns 5100 - 75000.
Ruger, Standard Auto with red eagle, pistol with S/Ns 1 - 25000.
Ruger, Super Black Hawk, revolvers having a barrel length of 6½ inches with S/Ns 24000 - 26000.
Ruger, Super Single Six stainless with 4" barrel, revolver with S/Ns 62-07500 - 64-650000.
Ruger, Super Single Six stainless with 9" barrel, revolver with S/Ns 62-07500 - 63-40000.
Ruger, Super Single Six chrome with 4-5/8" barrel, revolver with S/Ns 504000 - 505000.
Ruger, "21 Club" No. 1, rifle with random S/Ns.
Ruger, 44 Deerstalker, carbine with S/Ns 1 to 5000.
Ruger, S47 Code, Model Super Blackhawk, 7 ½ inch barrel, .44 magnum revolvers, in S/N range 196 to 3111, with long grip frame, micro rear sight, and in mahogany wood case, approximately 500 manufactured.
Rumanian Mosin Nagant 1944 type carbines, caliber 7.62 x 54R, manufactured in Romania, and manufactured from 1952 to 1956.
Russian (U.S.S.R.), Nagant revolver, model 1895, cal. 7.62 Nagant and .22 cal. all variations, mfd. by the Tula Arsenal, Tula, Russia, after 1898.
Russian (U.S.S.R.), Tokarev, model TT, 1930, pistol, cal. 7.62, mfd. at the Tula Arsenal, Tula, U.S.S.R., from 1939 through 1956.
Russian (U.S.S.R.), Tokarev, model TT, 1933, pistol, cal. 7.62, mfd. at the Tula Arsenal, Tula, U.S.S.R., from 1933 through 1956.
Russian (U.S.S.R.) Tokarev, model TT R-3, .22 cal., pistol.
Russian (U.S.S.R.) Tokarev, model TT R-4, .22 cal., pistol.
Russian (U.S.S.R.), Tula Korovin, Tk, .25 ACP cal., semiautomatic pistol.
Russian (U.S.S.R.), model 1891, Mosin-Nagant rifles, cal. 7.62 x 54R and .22 cal., all models and all variations, mfd. after 1898 (i.e., M1891/30, M1910, M1938, and M1944).
Russian (U.S.S.R.), Tokarev, semiautomatic rifle, model 1938 (SVT38), cal. 7.62 x 54R, of Soviet manufacture.
Russian (U.S.S.R.), Tokarev, semiautomatic rifle, model 1940 (SVT40), cal. 7.62 x 54R, of Soviet manufacture.
Russian (U.S.S.R.), Tokarev, semiautomatic carbine, model 1932 (nonstandard), cal. 7.62 x 54R, of Soviet manufacture.
Russian (U.S.S.R.), Tokarev, semiautomatic carbine, model 1940 (SVT40), cal. 7.62 x 54R, of Soviet manufacture.
Russian (U.S.S.R.), Simonov, semiautomatic rifle, model SKS, cal. 7.62 x 39, of Soviet manufacture.
Russian (U.S.S.R.), Dragunov, semiautomatic rifle, model SVD, cal. 7.62 x 54R, of Soviet manufacture, Soviet military issue only.
* Sako, Anniversary Model, 7mm Remington magnum caliber, bolt action rifle.
J. P. Sauer & Sohn pistols, mfd. prior to 1946.
Sauer, 38(h), pistol, cal. 7.65mm marked w/Third Reich police acceptance stamps of Eagle C, F, K, or L. Savage Arms, semiautomatic pistols, cal. .45 ACP, all models.
Savage Arms, model 99, lever action, centerfire rifles, mfd. in Utica, NY prior to World War II with S/Ns below 450000.
Savage, Prototype pistols, cal. .25, .32 and .38 made between 1907 and 1927.
Savage model 1907 pistol, caliber .32 and .380.
Savage model 1915 pistol, caliber .32 and .380.
Savage model 1917 pistol, caliber .32 and .380.
Schwarzlose, pocket model 1908 in 7.65mm, pistol mfd. by A.W. Schwarzlose, G.m.b.h., Berlin, Germany, and those assembled or made by Warner Arms.
Smith and Wesson, Model 624 revolver, First Issue, cal. .44 Target, engraved year "1985" over the issue number (1-25) on the right sideplate.
Smith & Wesson Collector's Association, 25th Anniversary Commemorative revolver 1970-1995, Model 29, .44 Magnum caliber, S/Ns SWC 0001 through SWC 0184.
Smith & Wesson, 125th anniversary Commemorative, model 25, revolver, cal. .45, marked "Smith & Wesson 125th Anniversary" and mfd. in 1977.
Smith & Wesson, 150th anniversary Texas Ranger Commemorative model 19 revolver.

Smith & Wesson, 1st model, Ladysmith revolver, cal. .22 rimfire long.
Smith & Wesson, .22/32 Kit Gun, cal. .22LR, S/Ns 525670 - 534636 (no letter).
Smith & Wesson, 2nd model, Ladysmith revolver, cal. .22 rimfire long.
Smith & Wesson, 2nd model, single shot pistol, cals. .22 rimfire, .32 S&W and .38 S&W.
Smith & Wesson, .32 Double Action Top Break, cal. .32 S&W, S/Ns 209302 and higher.
Smith & Wesson, .32 Safety Hammerless Top Break (New Departure), cal. .32 S&W, S/Ns 91401 and higher.
Smith & Wesson, .357 Magnum Hand Ejector, cal. .357 Magnum, S/Ns 45768 to 60000 (no letter).
Smith & Wesson, .38 Double Action Top Break Perfected model, cal. .38 S&W.
Smith & Wesson, .38 Double Action Top Break, cal. .38 S&W, S/Ns 382023 and higher.
Smith & Wesson, .38 Hand Ejector Military and Police, cal. .38, S/Ns 1 to 241703 (no letter).
Smith & Wesson, .38 Safety Hammerless Top Break (New Departure), cal. .38 S&W, S/Ns 19901 and higher.
Smith & Wesson, .38/44 Outdoorsman & Heavy Duty, cal. .38, S/Ns 36500-62023 (no letter).
Smith & Wesson, 3rd model, Ladysmith revolver, cal. .22 rimfire long.
Smith & Wesson, 3rd model, single shot pistol, cals. .22 rimfire, .32 S&W and .38 S&W.
Smith & Wesson, 4 screw side plate revolvers, old style N-frame series, with no model designation stamped in the yoke cut, in cal. .44 magnum, all barrel lengths, falling within the S130000-Sl60350 block of S/Ns, of which a total of 6,500 units were produced from 1956 to 1958.
Smith & Wesson, .44 Hand Ejector, all cal., S/Ns 1-62488 (no letter).
Smith & Wesson, .455 Mark II Hand Ejector, caliber .455.
Smith & Wesson, California Highway Patrol Commemorative model 19 revolver, cal. .357.
Smith & Wesson, City of Los Angeles 200th Anniversary Commemorative model 19 revolver, cal. .357.
Smith & Wesson, K-22 Hand Ejector, cal. .22 LR, S/Ns 632132-696952 (no letter).
Smith & Wesson, K-32 Hand Ejector (K-32 Masterpiece), cal. .32 S&W Long, S/Ns 653388 to 682207 (no letter).
Smith & Wesson, Mercox Dart Gun, cal. .22 rimfire, blank.
Smith & Wesson, Model 16 (K-32 Masterpiece), cal. .32 S&W Long, "K" S/N series.
Smith & Wesson, model 21, .44 Special caliber, also known as the".44 Hand Ejector Fourth Model" and the "1950 Model Military," having S/Ns S75,000 - S263,000.
Smith & Wesson, Model .22/32 Hand Ejector (Bekeart model), cal. .22LR, S/Ns 138220 to 534636 (no letter).
Smith & Wesson, Model 29 "Elmer Keith Commemorative" .44 magnum revolvers, S/N EMK1 – EMK2500.
Smith & Wesson, Model 39, Connecticut State Police 75th Anniversary 1903-1978, pistols, S/Ns CSP001 – CSP704.
Smith & Wesson, Model 39, steel frame pistol, cal. 9mm parabellum.
Smith & Wesson, Model 39-1 (52-A), pistol, cal. 9mm parabellum.
Smith & Wesson, Model 53, Remington Jet Center Fire Magnum, cal. .22.
Smith & Wesson, Model 41-1 .22 short cal. semiautomatic pistols.
Smith & Wesson, Model 42 Centennial Airweight .38 Special 5-shot revolvers, with aluminum alloy frames and cylinders.
Smith & Wesson, Model 45 Military & Police .22LR cal. revolvers.
Smith & Wesson, Model 46 .22LR caliber semiautomatic pistols.
Smith & Wesson, Model 56 U.S. Air Force contract .38 Special 6-shot revolvers, S/N K500001 through K515001.
* Smith & Wesson Model 66 Bureau of Alcohol, Tobacco & Firearms 1933 – 1983 50th Anniversary Commemorative .357 magnum caliber revolvers, having a "BATF" prefix serial number and miniature replica of an agent's badge engraved into the frame.
Smith & Wesson Model 66 Distinguished Combat Magnum, caliber .357 magnum revolver, marked with the Texas Sheriff's Association badge and "TEXAS LAWMAN" on the right side of the frame, commemorating 150 years of law enforcement in Texas 1836-1986.
Smith & Wesson model 66, Dallas Police Department Commemorative Edition 1881-1981.
Smith & Wesson, Model 66, "Naval Investigative Service Commemorative" .357 6-shot revolvers.
Smith & Wesson, Model 147-A, 9mm 14-shot semiautomatic pistols.
Smith & Wesson, Model 544, "Texas Wagon Train Commemorative" .44/40 cal. 6-shot revolvers,

S/N TWT001 through TWT7800.
Smith & Wesson, .45 hand ejector model of 1950 Military, .45 caliber, having S/Ns between S76000 and S263000.
Smith & Wesson, Model 1917 revolver, cal. .45 ACP, produced for Brazil.
Smith & Wesson, Model Straight Line, single shot pistol, cal. .22 rimfire long rifle.
Smith & Wesson, pistol, caliber .32 ACP.
Smith & Wesson, pistol, caliber .35, all variations.
Smith & Wesson Registered Model 27 revolvers, cal. .357-magnum, 50 Yr. Commemorative, 5" barrel, S/Ns REG0001 through REG2500, inclusive.
Smith & Wesson U.S. Air Force contract M13 Aircrewman .38 Special 5-shot J frame and 6-shot K frame revolvers.
Smith & Wesson, U.S. Border Patrol 50th Anniversary Commemorative, model 66, stainless steel, cal. .357 Magnum, revolvers.
Sosso, pistols, mfd. by Guilio Sosso, Turin, Italy, or Fabrica Nationale D'Armi, Brescia, Italy, cal. 9mm.
Springfield Armory, Inc., Korean War Commemorative .30 cal., MI Garand Rifle S/Ns from KW0001 to KW1000.
Standard Arms Co., rifle/shotgun combination, U.S., model "Camp," slide action cal. .50.
Standard Arms Co., rifle model G, slide action or gas operated, any caliber.
Standard Arms Co., rifle model M, slide action caliber .25-.35, .30 Rem. and .35 Rem.
* Stevens, Models 425, 430, 435, and 440 High-Power Repeating Rifles in calibers .25 Remington, .30 Remington, .32 Remington, and .35 Remington.
Stevens, Model 77E, 12 gauge, military riot type shotguns, properly marked and identified as mfd. for the U.S. Military, from 1963 to 1969, in original configuration.
Steyr, model 1909, .25 ACP cal. semiautomatic pistol.
Steyr-Hahn, M1912, pistol, cal. 9mm Steyr.
Steyr-Hahn, M1912, pistol, cal. 9mm parabellum marked with Third Reich police acceptance stamps of Eagle C, F, K, or L.
Stock, semiautomatic pistols, all cals. mfd. by Franz Stock.
Swiss Model 1931/55 rifles, cal. 7.5 mm, S/Ns 1001 to 5150.
Swiss Schmidt Rubin, Model 1911, rifle made into a harpoon gun, caliber 12mm.
Swiss self loading rifle, test 1947, cal. 7.5mm, all variations.
Tauler, model military and police pistol.
Thompson/Center 25th Anniversary Contender, cal. .22LR pistol, 10 inch barrel, S/Ns 25 001-25 536.
Thompson Center Contender Pistol, cal. .30 Herrett, Steve Herrett Commemorative, S/Ns SH-001 to SH-500.
Thompson Center Contender Pistol, cal. 7mm TCU, IHMSA 10th anniversary commemorative, S/Ns IHMSA 10-001 to IHMSA 10-200.
Tippman Arms Company Models 1919 A-4, 1917, and .50 HB ½ scale, .22 caliber semi-automatic firearms, manufactured in Fort Wayne, Indiana, from 1986-1987.
Tokagypt 58, pistol, caliber 9mm parabellum.
Trejo, semiautomatic pistols, cal. .22, .32, and .380. mfd. in Mexico, circa 1952 to 1972.
* Uberti, 125th Anniversary of the Winchester, Model 1873 lever action rifle, .44/40 caliber, having S/Ns between A125th001 through A125th125.
Uberti Model 1866 lever action rifles, "An Engraver's Tribute to Gustave Young," .44/40 caliber,S/Ns GY001 through GY300.
Uberti model 1866 L.D. Nimschke, lever action rifles, caliber .44-40, S/Ns LDN001 through LDN300.
Uberti, Single action revolver, cal. .45, General George S. Patton Commemorative, S/Ns P0001 to P2500.
United States Patent Fire Arms Manufacturing Company, Artillery Model single action revolver, 100th anniversary of The Charge Up San Juan Hill, .45 caliber, with a 5 ½ - inch barrel and having S/Ns from NFM001 through NFM500 U.S, model 1911-A-1, .45 cal. pistol, mfd. by Union Switch and Signal Company, prototype model, with S/Ns US&S Exp. 1 to US&S Exp. 100.
U.S., model 1911-A1, semiautomatic, pistol, cal. .45, mfd. by the Singer Manufacturing Company in 1942, S/N range from S800001 to S800500.

U.S., model 1911-A1, semiautomatic pistol, cal. .45, mfd. by Remington Rand, bearing S/N prefix of ERRS.
U.S., model 1911-A1 semiautomatic pistol, cal. .45, produced as original factory cut-a-ways.
U.S., Rifle, cal. .30 M1, original military issue only, produced prior to 1956.
U.S. Rifle, caliber .30, M1, original issue only, produced prior to 1958.
U.S., Rifle, cal. .30, MC-1952, equipped with telescopic sight mount MC, telescopic sight MC1, marked U.S.M.C. or Kollmorgan.
U.S. Repeating Arms, Wyoming Centennial, Winchester Commemorative Model 94 Carbines, cal. .30-30, S/Ns WYC001 through WYC500.
UZI, Model A, semiautomatic carbine, cal. 9mm, having a satin nickel finish applied at the factory, S/Ns SA 0001 to SA 0100.
Walther, bolt action and semiautomatic rifles, all cals., mfd. prior to 1946.
Walther, model PP and PPK semiautomatic pistols, in all cals., mfd. in France and marked "MANURHIN".
Walther, Model PP pistol, 50 Jahre 1929-1959 Commemorative, caliber .22 and .380.
Walther, Model TP and TPH pistols, cal. .22 and .25 ACP, original German manufacture only.
Walther, Olympic bolt action single shot match rifle, in cal. .22 made by Waffenfabrik Walther, Zella-Mehlis (thur.) prior to World War II.
Walther, pistols, mfd. at Zel1a-Mehlis prior to 1946, all models, any caliber.
Walther, rifles, model 182, cal. .22 made by Waffenfabrik Walther, Zelia-Mehlis (thur.) prior to World War II.
Watson Brothers, Lee-Speed type, custom sporting rifle, cal. .303, S/N 7738.
Webley and Scott, model 1910 and 1913 high velocity pistols, cal. .38 ACP.
Webley and Scott, M1913, Navy or Commercial, self loading pistol, cal. .455.
Webley-Fosbury, semiautomatic revolvers, all cals., all models.
Webley, model 1909, pistol, cal. 9mm Browning Long.
Whitney, "Wolverine" and "Lighting" .22 cal. automatic pistols as mfd. by Whitney Firearms Company, Hartford, CT between 1955 - 1962.
Winchester, Model 1894, Florida Sesquicentennial carbine, cal. .30-30, S/Ns FL001-FL500.
Winchester, Apache Commemorative carbine, commemorative edition of model 1894 Winchester with S/N prefix of A.
Winchester, Comanche Commemorative carbine, commemorative edition of model 1894 Winchester with S/N prefix of CC.
Winchester, "Ducks Unlimited" shotgun, model 12, bearing S/Ns DU-001 through DU-800 (Commemorative).
Winchester, Kentucky Bicentennial Model 94 carbines, cal. .30-30, S/N range KY001-KY500.
Winchester, "Matched Set of 1000," a cased pair consisting of a Winchester model 94 rifle, cal. .30-30 and a Winchester model 9422 rifle, cal. .22.
Winchester, Model 12, pump action shotguns, mfd. from 1912 through 1963, S/Ns 1 through 1962017.
Winchester, Model 12, Shotgun, 12 gauge, prototype, for Ducks Unlimited Commemorative, S/N Y2002214.
* Winchester, Model 12 Y, pump action shotguns, all variations, manufactured between 1972 and 1979, S/Ns Y2000100 through Y2026399.
Winchester, Model 21, double barrel shotguns, all gauges, all grades, mfd. by Winchester and U.S. Repeating Arms.
Winchester, Model 21, Grand American Double Barrel Shotgun, cal. 20 and 28 gauge, S/N 32984, Engraved Custom Built by Winchester for Philip S. Rane.
Winchester, Model 37, single barrel shotguns, all gauges, manufactured between 1936 and 1963.
Winchester, Model 42, .410 gauge shotguns.
Winchester, Model 52, rifle, bearing S/Ns 1 to 6500.
Winchester, Model 53, all original, mfd. from 1924 to 1947 with 16" or longer barrel, and 26" or longer overall length.
Winchester, Model 54 rifles.
Winchester, Model 54, rifle, speed lock variation, cal. .270.
Winchester, Model 55, .22 caliber, single shot rifle.
Winchester, Model 61, cal. .22 rimfire, slide action repeater, hammerless.

Winchester, Model 62, cal. .22 rimfire, slide action repeater.
Winchester, Model 63, self loading rifles, cal. .22 rimfire.
Winchester, Model 64 and 65, lever action rifles.
* Winchester, Model 70, .300 Winchester Magnum, 50th Anniversary bolt action rifle, S/Ns 50 ANV 1 through 50 ANV 500.
Winchester, Model 70, bolt action rifle, all cals., mfd. in or before 1963 having S/Ns less than 581472.
Winchester, Model 70, rifle, cal. .308 rifle, 19" barrel and Mannlicher type stock, made from 1968 to 1971.
Winchester, Model 70, rifles, .308, .270 Winchester, and 30-06 cal., 19" barrel and Mannlicher type stock, made from 1968 to 1971.
Winchester, Model 70, Ultra Match Target Special Grade rifle, cal. .308.
* Winchester, Model 70, XTR Featherweight Ultra Grade, .270 caliber, bolt action rifle, 1 of 1,000.
Winchester, Model 71, all original, mfd. from 1936 to 1958 with 16" or longer barrel and 26" or longer overall length.
Winchester, Model 85, (single shot rifle), all original, mfd. from 1899 to 1920, with 16" or longer barrel and 26" or longer overall length.
Winchester, Model 86, all original, mfd. from 1899 to 1935, with 16" or longer barrel and 26" or longer overall length.
Winchester, Model 88, carbine, cal. .243, .284, .308, or .358 mfd. by Winchester Western Division, Olin Corporation, New Haven, CT.
Winchester, Model 88 rifles, all calibers.
Winchester, Model 92, all original, mfd. from 1899 to 1947, with 16" or longer barrel and 26" or longer overall length.
Winchester, Model 94, Alaskan Purchase Centennial, carbine.
Winchester, Model 94, American Bald Eagle Commemorative Carbine.
Winchester, Model 94, Arapaho Commemorative carbines, caliber .30-30, S/Ns ARAPA001 – ARAPA500.
Winchester, Model 94, Bat Masterson commemorative.
Winchester, Model 94, Bicentennial 76, carbine.
Winchester, Model 94, Buffalo Bill, carbine.
Winchester, Model 94, Buffalo Bill, rifle.
Winchester, Model 94, C.M. Russell, Great Western Artist Commemorative Carbine.
Winchester, Model 94, cal. .30-30, Antlered Game Commemorative, carbine.
Winchester, Model 94, cal. .30-30, Legendary Lawman Commemorative, carbine.
Winchester, Model 94, Calgary Stampede Commemorative, carbine, cal. .32 Winchester Special.
Winchester, Model 94, Canadian 1967, Centennial carbine.
Winchester, Model 94, Canadian 1967, Centennial rifle.
Winchester, Model 94, carbine, Canadian Pacific Centennial, cal. .32 Winchester Special.
Winchester, Model 94, carbine, Oklahoma Diamond Jubilee Commemorative.
Winchester, Model 94, carbines, Chevrolet Outdoorsman sets, .30-30 cal., having S/Ns between 5130000 and 5466000.
Winchester, Model 94, "Chief Crazy Horse," commemorative lever action rifle, cal. .38-55, mfd. by U.S. Repeating Arms Co., New Haven, CT.
Winchester, Model 94, Colt Commemorative Set, Winchester Signature model "Oliver F. Winchester" carbine, cal. 44-40, Lever Action, as mfd. by U.S. Repeating Arms Co., New Haven, CT.
Winchester, Model 94, Cowboy Commemorative, carbine.
Winchester, Model 94 rifle, Custom Limited Edition Centennial, .30 WCF cal., S/Ns CNTL 01 through CNTL 94.
Winchester, Model 94, Frederick Remington, Great Western Artist Commemorative Carbine.
Winchester, Model 94, Illinois Sesquicentennial, carbine.
Winchester, Model 94, in cal. .38/55, 1980, Alberta Diamond Jubilee Commemorative, carbine.
Winchester, Model 94, John Wayne Commemorative (Canadian Issue), carbine, cal. .32-40.
Winchester, Model 94, John Wayne commemorative, cal. .32-40, carbine.
Winchester, Model 94, Klondike Gold Rush Commemorative carbine.
Winchester, Model 94, Legendary Frontiersman rifle, cal. .38-55.
Winchester, Model 94, "Limited Edition I."
Winchester, Model 94, "Limited Edition II" rifle, cal. .30-30.

Winchester, Model 94, Limited Edition, carbine, cal. .30-30, S/Ns 77L1 - 77L1500.
Winchester, Model 94, Limited Edition Centennial, .30 WCF rifles, S/Ns CN10,001 - CN10,250, mfd. by the USRAC Custom Gun Shop for the Winchester Arms Collectors Association, having a replica Lyman No. 2 tang sight and with the WACA medallion placed in the buttstock.
Winchester, Model 94, Little Big Horn Centennial, carbine.
Winchester, Model 94, Lone Star Commemorative, carbine.
Winchester, Model 94, Lone Star Commemorative, rifle, .30-30.
Winchester, Model 94, model NRA Centennial, carbine.
Winchester, Model 94, Mounted Police, carbine.
Winchester, Model 94, Nebraska Centennial, carbine.
Winchester, Model 94, NRA Centennial rifle, .30-30.
Winchester, Model 94, One of One Thousand European Rifle commemorative.
Winchester, Model 94, Ontario, Canada Conservation Office, 100th Anniversary Commemorative carbines, caliber 30-30.
Winchester, Model 94, rifles and carbines mfd. prior to January 2, 1964, and having a S/N of less than 2,700,000, provided their barrel length is at least 16" and their overall length at least 26".
Winchester, Model 94, Royal Canadian Mounted Police Centennial carbine.
Winchester, Model 94, Saskatchewan Diamond Jubilee Carbine commemorative.
Winchester, Model 94, Sioux Commemorative carbines, cal. 30-30, with S/Ns beginning with "SU."
Winchester, Model 94, The Oliver F. Winchester commemorative.
Winchester, Model 94, Theodore Roosevelt, Carbine.
Winchester, Model 94, Theodore Roosevelt, Rifle.
Winchester, Model 94 "Trapper," .357 magnum caliber saddle ring carbine, 100th Anniversary Commemorative 1894-1994Winchester, Model 94, United States Border Patrol Commemorative carbine cal. .30-30.
Winchester, Model 94, Wild Bill Hickok Commemorative, Caliber .45, S/Ns WBH001 - WBH350.
Winchester, Model 94, Wells Fargo and Company Commemorative, carbines.
Winchester, Model 94, Wyoming Centennial Commemorative carbines, caliber .30-30, S/Ns WYC001 - WYC500.
Winchester, Model 94, Wyoming Diamond Jubilee, carbine.
Winchester, Model 94, Yellow Boy Indian, carbine.
Winchester, Model 94, 125th Anniversary Commemorative carbines, caliber .30-30, S/Ns WRAC001 - WRAC125.
Winchester, Model 94, 150th Anniversary Texas Ranger Commemorative, carbine.
Winchester, Model 95, all original, mfd. from 1899 to 1938, with 16" or longer barrel and 26" or longer overall length.
Winchester, Model 1866, Centennial, carbine.
Winchester, Model 1866, Centennial, rifle.
Winchester, Model 1873, all original, mfd. from 1899 to 1925, with 16" or longer barrel and 26" or longer overall length.
Winchester Model 1890 rifles.
Winchester, Model 1892 Carbine, .38 WCF cal. with 14" barrel S/N 654311.
Winchester, Model 1894, Golden Spike, carbine.
Winchester, Model 1894 Nez Perce Commemorative carbine, cal. .30-30, mfd. by U.S. Repeating Arms Co., S/Ns NEZ001-NEZ600.
Winchester, Model 1894, Texas Sesquicentennial rifle and carbine, cal. .38-55.
Winchester, Model 1894 Carbine, .30 WCF cal. with 15 1/16" barrel S/N 1040988.
Winchester, Model 1894 Special Carbine .30 WCF cal. S/N 948487.
Winchester, Model 1897, pump action shotguns, all gauges, manufactured between 1899 and 1957, S/Ns 1 through 1,024,700.
Winchester, Model 1897 or 97 riot guns, 12 gauge, 20" barrels with original RIC (Royal Irish Constabulary) markings.
* Winchester, Model 1902, .22 caliber bolt action rifles.
Winchester, Model 1903 .22 cal. semiautomatic rifles.
Winchester, Model 1906 rifles.
Winchester, Model 9422, Annie Oakley Commemorative Carbine .22 cal.

SECTION II

Winchester, Model 9422, Cheyenne Commemorative rifles, .22 cal.
Winchester, Model 9422, "Eagle Scout Commemorative" rifle, .22 cal.
Winchester, Northwest Territories Centennial rifle.
Yugoslavian M1948 (M48) bolt action rifles, 7.92 x 57mm caliber, produced at the Kragujevac Arsenal, original military configuration only.
Yugoslavian manufactured rifles M59 and M59/66, 7.62 x 39 mm caliber, all semiautomatic variations and having a fixed magazine, manufactured from 1947-1992.

SECTION III: Weapons Removed From The NFA As Collector's Items And Classified As Curios Or Relics Under The GCA

The Bureau has determined that by reason of the date of their manufacture, value, design and other characteristics, the following firearms are primarily collector's items and are not likely to be used as weapons and, therefore, are excluded from the provisions of the National Firearms Act.

Further, the Bureau has determined that such firearms are also curios or relics as defined in 27 CFR 478.11. Thus, licensed collectors may acquire, hold, or dispose to them as curios or relics subject to the provisions of 18 U.S.C. Chapter 44 and 27 CFR Part 478. They are still "firearms" as defined in 18 U.S.C. Chapter 44.

Any military bolt action or semiautomatic rifle mfd. prior to 1946 and accompanied by a "cup type" grenade launcher designed for the specific rifle.

Belgian, Pre-war mfd. Hi Power pistols, in cal. 9mm having tangent sights graduated to 500 meters, slotted for shoulder stock, having S/Ns of less than 47,000 without letter prefixes or suffixes and accompanied by original Belgian mfd. detachable wooden flat board type shoulder stocks.

Beretta, model 1918/1930, semiautomatic carbine, cal. 9mm, having a barrel length of 12.5" and a magazine capacity of 25 rounds.

Beretta, model 1923, semiautomatic pistol, in cal. 9mm Kurz (.380), accompanied by original Italian detachable leather and metal holster/shoulder stock.

Bergmann-Bayard, model 1908, Pistol, 9mm Bergmann-Bayard with shoulder stock and 4" barrel.

Bergmann, Mars model 1903, self loading pistol, w/accompanying shoulder stock.

British PIAT (Projector, Infantry, Anti-tank).

Browning, model 1903, Pistol, 9mm Browning Long, with shoulder stock and 5" barrel.

Browning Hi power pistols, 9mm having tangent sights graduated to 500 meters, slotted for shoulder stock, having S/Ns less than T200,000 etched vertically on the right side of slide, barrel, or frame and bearing crest of Emirates of Muscat & Oman, or mirror image of such crest, accompanied by original detachable wooden flat board shoulder stocks.

Canadian, Inglis No. 1, Chinese Contract, Hi Power pistols, cal. 9mm parabellum, having a tangent rear sight adjustable from 50 to 500 meters, slotted for shoulder stock, and having the letters C in the S/N and accompanied by original Canadian mfd. detachable wooden holster/shoulder stock.

Clement, Pistol Carbine, cal. 9mm.

Chinese, Jingal rifle, caliber .60, S/N IRS 9083H.

Chinese mfd. copies of the Mauser model 1896, semiautomatic pistol, produced prior to 1945, any cal., accompanied by original Chinese mfd. detachable wooden holster/shoulder stocks.

Colt, Lightning rifle, original cutaway demonstrator produced by Colt.

Colt, Model New Service, cal. .44 WCF, having a smooth bore, S/Ns 309065, 316627.

Colt, Model New Service revolver, cal. .45 Colt, S/N 326618, having a smooth bore.

Colt, Model New Service revolver, caliber .44 WCF, S/N 55125, having a barrel with a smooth bore.

Colt, Model Woodsman, .22 Long Rifle cal., semiautomatic pistol with an experimental 10" barrel and an experimental wooden detachable shoulder stock.

Colt, Model Woodsman, .22 Long Rifle cal., semiautomatic pistols, mfd. between 1915 and 1943, together with the original leather detachable holster stocks, mfd. by the N&S Corporation, Ventura, CA.

Colt, Model 1905, Pistol, .45 rimless, with leather holster/shoulder stock and 5" barrel.

Colt, Model 1911 pistol, cal. .38 Super, S/N 16212, accompanied by an original "Multi-Shot Riot and Anti-Bandit" shoulder stock as originally mfd. by Monarch Arms Sales Company, Los Angeles, California, in the early 1930s.

Colt, New Service Revolver, cal. 44/40, w/original smooth bore barrel, S/N 326596.

Colt, Officers model, .38 Special cal., double action revolver, with 6" barrel and a detachable, experimental skeleton shoulder stock and holster combination.

Colt, semiautomatic pistol, cal. .22 rimfire, S/Ns 49551 and 49712, accompanied by an original "Rifle-ette" shoulder stock as originally mfd. by Monarch Arms & Mfg Co., Los Angles,

CA in the 1920s.

Colt, Single Action Army revolver, cal. .45, with original smoothbore barrel, S/N 325085.

Colt, Single Action Army revolver, caliber 44 WCF (.44-40), S/N 324477, with blue finish, 5-1/2 inch barrel and smooth bore presented by Colt Firearms to Mr. John W. Garrett.

Colt, Single Action Army revolver, S/N 354096, cal. .44/40 having a smooth bored barrel and a barrel length of 7-1/2".

Czechoslovakian, model CZ24, semiautomatic pistol, in cal. 9mm Kurz (.380) accompanied by original Czechoslovakian detachable wooden holster/shoulder stock.

Fiala, model 1920, repeating pistol, cal. .22LR in all barrel lengths with accompanying detachable shoulder stock; original copies of the Fiala repeating pistol, marked Schall, Columbia or Botwinick Brothers, cal. .22LR, with accompanying original detachable shoulder stock.

Finnish, model L-35, Lahti, Semiautomatic pistol, in cal. 9mm parabellum, accompanied by original Finnish detachable wooden holster/shoulder stock.

Flashlight Revolver, cal. .22 Rimfire, having a rifled barrel, patented on July 10, 1923, and originally mfd. by S. P. Cottrell in Buffalo, NY.

Fromme, model 1912, semiautomatic pistol w/Benke-Thiemann folding shoulder stock.

German, model 1918, (WWI) anti-tank rifle (PzAgew 1918), cal. 13.25mm.

German, (Nazi) Belt Buckle Gun, .22 rimfire, marked "DRP Ausl Pat, Louis Marquis, W. Elberfeld."

German, (Nazi) Belt Buckle Gun, 7.65mm, marked "D.R.P. Angem."

German, Kamphpistole, cal. 26.7mm.

German, Leuchpistole, 26.7mm, Walther pattern, mfd. in or before 1945, with original 23mm rifle grenade launching adapter sleeve.

German, VG1-5 (Volksgewehr), semiautomatic rifle, cal. 7.92mm Kurz, having a barrel length of 14.9" and an overall length of 34.8".

Greener Cattle Killer (Original model) No. B1201, .310 cal.

Greener Cattle Killer (Pocket Pattern) No. B1203, .310 cal.

Greener Safti Killer No. B1216, .22 cal.

Greener Universal Safti Killer No. B1217, .310 cal.

Hamilton, models 7, 11, 15, 19, 23, 27, 027, 31, 35, 39, and 43 rifles.

Harrington and Richardson (H&R) "Handy Gun," .22 rimfire or .32-20 centerfire, with 12-1/4" rifled barrel, without shoulder stock.

Harrington and Richardson Handy Gun, Model 122 .22 Rimfire caliber, with original shoulder stock, having a barrel length of 12-3/4 inches, S/N 157.

Heal, Rifle No. 10, cal. .22 Heal, .22 rimfire cal. rifles, all models, mfd. prior to 1908, by the Heal Rifle Company or the Detroit Rifle Company.

High Standard, model C/S, smoothbore .22 cal. shot semiautomatic pistols, bearing S/Ns 59279, 59460, 59469, 59473, 59478.

High Standard, Model Sentinel (R-100), cal. .22 revolver having smooth bore with integral choke, S/Ns 948, 995.

High Standard, Model S, smoothbore .22 cal. shot semiautomatic pistols having slides marked "Hi-STANDARD MODEL "S" .22 L.R. SHOT ONLY," and bearing S/Ns 48142, 48143, 48144, 48145, 48146, 59474, 59496, 59458, 59459.

High Standard, Model S, smoothbore pistol, caliber .22 L.R., S/N 59493.

* Invisible Defender four barrelled palm gun in .22 caliber manufactured by O.F. Mossberg, Chicopee Falls, Massachusetts, circa 1902 – 1909.

Japanese, 20mm Single-Barrel Variant, flare pistol.

"JGA" (J.G. Anchutz, Ulm, Germany), .22 Flobert single shot pistol.

J.G. Anschutz, bolt action pistol, single shot, cal. 6 mm Glatt, S/N 377153.

Kennedy Repeating Rifle, cal. .44, with 14" barrel, S/N C329.

Krupp, models 1902 and 1906, 50mm Mountain Cannons as produced for the Siamese Government.

Luger, Artillery model, pistols having chamber dates of 1914 through 1918 or 1920, having German Weimar Navy markings consisting of the letter M over an anchor and a German Navy property number accompanied by original Artillery Luger flat board stocks, bearing German Weimar Navy markings of the letter M over an anchor with or without Navy property numbers.

Luger, the 1920 Commercial Artillery model, pistols as mfd. by DWM or Erfurt, having undated chambers, commercial proofmarks, and bearing the inscription Germany or Made in Germany on the receiver and accompanied by original, German mfd., artillery type, detachable wooden shoulder stocks.

Luger, DWM Pistol, model 1900, 1902, or 1906, in 7.65 Luger or 9mm parabellum cal., having the American Eagle chamber crest, and barrel lengths of either 4" or 4-3/4", with original detachable Ideal shoulder stocks and Ideal frame grips.

DWM Luger, Original models 1904, 1906, 1908, 1914, and 1920. Naval pistols in 9mm parabellum or 7.65mm cal., in both the Commercial and Naval military varieties; in both altered and unaltered barrel lengths in the model 1904 and in both altered and unaltered safety markings in the model 1906; with original board-type detachable shoulder stocks bearing brass or iron discs, with or without markings, or, if without brass or iron discs, being of the Navy flat board-type. This exemption applies only to the listed Naval Luger pistols if mated to the Naval Luger stock and will not apply if the Naval Luger pistol is mated to the Artillery stock. The Naval stock has an overall dimension of 12-3/4", a rear width of 4-5/8", a front width of 1-1/2", a rear thickness of 9/16", and a front thickness of 1-3/16".

Luger, DWM Stoeger model 1920 and 1923, semiautomatic pistols in 7.65mm or 9mm parabellum cal., in barrel lengths of 8, 10, 12, and 12-1/2", having either American Eagle chamber crests and/or Stoeger frame and/or upper receiver marks, having either standard, Navy or artillery rear sights, having extractors marked either "Loaded" or "Geladen" and having frame safety markings of either "Gesichert" or "Safe," together w/original commercial flat board stocks of the artillery type, which bear no S/Ns or military proof marks; may include a "Germany" marking.

Luger, DWM Pistol-Carbine, model 1920, 7.65mm or 9mm parabellum cal., with accompanying original commercial type shoulder stock, with or without forearm piece, having barrel lengths of 11-3/4" to less than 16".

Luger, German model 1914, Artillery model pistol, mfd. by DWM or Erfurt, having chambers dated 1914 –1918, bearing Imperial German military proofmarks & accompanied by original, German mfd., artillery type, detachable wooden shoulder stocks.

Luger, model 1902, Pistol-Carbine, 7.65mm Luger with original commercial type shoulder stock and forearm and 11-3/4" barrel.

Luger, Persian (Iranian) Artillery model, pistols, as mfd. by Mauser prior to 1945, accompanied by the original artillery type, detachable wooden shoulder stock, bearing a S/N in Farsi characters stamped into the wood on the left side.

Luger, semiautomatic pistol, certain variations with Benke-Thiemann folding shoulder stock.

MBA, Gyrojet Rocket Guns, cal. 13mm, semiautomatic version only, produced in 1968 or earlier, S/N ranges A0001-A0032, A001-A085, B010-B411, and B5059-B5692.

Manville, 18-shot-drum, 25mm, semiautomatic tear gas gun.

Marlin, Model 1892, caliber .44-.40, S/N 386485, with 15-inch barrel.

Marlin Model 1893 or 93 carbine, cal. .25, with 15 inch barrel, S/Ns 193134, 284957, 320283, 387186.

Marlin Model 1893 or 93 carbine, cal. .30, with 15 inch barrel, S/Ns 199669, 199690, 203990, 204585, 208526, 228856, 237681, 244372, 256559, 256568, 256577, 256583, 256611, 258853, 261844, 261889, 261901, 269397, 269636, 281802, 281810, 291358, 336467, 336478, 336504, 336610, 338883.

Marlin Model 1893 carbine, .30-30 caliber having a factory fitted 15-inch barrel S/N 1934.

Marlin Model 1893 carbine, cal. .30-30, with a 15 inch barrel, S/Ns 244371, C2231 and 269397.

Marlin Model 1893, carbine, cal. .32 with 15 inch barrel, S/N C2315.

Marlin Model 1893 or 93 carbine, cal. .32-30, .32-40, .32 HPS, with 15 inch barrel, S/Ns *2293, 176656, 194471, 194786, 194771, 194786, *251864, 258880, 258894, 285305, 285331, 291337, 291399, 292390, 307905, 308094, 308111, 329063, 335434, 336651, 336964, 370263, 405878, 426311, C1326, C1323, C3967.

Marlin Model 1893 or 93 carbine, cal. .38-55 or .38 HPS, w/15 inch barrel, S/Ns 194783, 246363, 75896, 316309, 336529, A4054, C3394, D8266.

Marlin Model 1894 or 94 carbine, cal. .25, with 12 inch barrel, S/Ns 283149, 328901, 328925, 328928, 328937, 328958.

Marlin Model 1894 or 94 carbine, cal. .25, with 15 inch barrel, S/Ns 190958, 190962, 190976, 211087, 280125, 280128, 280141, 280147, 283330, 287748, 294716, 328910, 337955, 442611, 442643.

SECTION III

Marlin Model 1894, .25-20 caliber, S/Ns 409198, 442606, having a 15-inch barrel.

Marlin Model 1894 or 94 carbine, cal. .32, with 15 inch barrel, S/Ns 190944, 190950, 190963, 190972, 217948, 282898, 282904, 287741, 287745, 287753, 287755, 287762, 287764, 287769, 287770, 287772, 294666, 294732, 301560, 314652, 325609, 328807, 328837, 328933, 376900.

Marlin Model 1894, caliber .32-20, S/N 435220, with 15-inch barrel.

Marlin Model 1894 carbine, caliber .33-20, S/Ns *D8959 and 417837, having a 15-inch barrel.

Marlin Model 1894 or 94 carbine, cal. .38, with 14 inch barrel, S/Ns 324849, 326718.

Marlin Model 1894 or 94 carbine, cal. .38, with 15 inch barrel, S/Ns 195570, 199975, 200007, 20022, 215011, 216362, 216412, 282004, 282936, 282997, 285870, 285974, 285979, 285990, 286281, 287729, 300626, 310864, 316798, 318000, 321957, 321970, 321973, 321985, 321986, 321990, 322025, 322035, 322037, 322053, 322077, 322079, 322082, 322095, 322514, 323672, 323688, 323785, 323810, 323846, 323925, 323928, 324195, 324238, 325732, 325740, 326143, 326206, 326726, 330543, 384186, 397215.

Marlin Model 1894 or 94, caliber .38-40, S/N 430591, as originally manufactured with a 15-1/4 inch barrel.

Marlin Model 1894 or 94 carbine, cal. 44, with 14 inch barrel, S/Ns 312691, 312708, 312801, 316617, 316628, 316632, 316645, 316649, 316654, 316656, 316739, 316772, 316803, 317427, 317449, 317466, 317470, 317499, 317559, 317573, 317578, 317582, 317587, 317596, 317646, 317687, 317690, 317698, 317697, 322641.

Marlin Model 1894 or 94 carbine, cal. .44, with 15 inch barrel, S/Ns 181852, 181863, 181911, 181967, 182048, 182068, 182093, 182205, 182219, 182226, 182321, 182322, 182344, 183374, 183376, 183379, 183400, 183401, 183403, 183408, 183413, 183418, 183428, 183432, 183433, 185128, 185129, 185133, 185145, 185150, 185155, 185165, 185167, 185174, 185175, 185185, 85186, 185187, 185192, 191033, 191096, 191097, 191098, 191099, 191100, 191101, 191104, 191105, 191106, 191424, 195823, 195840, 195842, 196827, 197867, 200830, 200903, 200906, 200922, 200928, 200929, 200930, 200958, 203760, 203761,203763, 203766, 203767, 203769, 203771, 203772, 203773, 203774, 203778, 203780, 203781, 203782, 203784, 203785, 203787, 203788, 203789, 203796, 203798, 203806, 203811, 203812, 203813, 203818, 203822, 203824, 203826, 204395, 204398, 204403, 204404, 204407, 204411, 204414, 204415, 204419, 204421, 204426, 204427, 204428, 204429, 204430, 204499, 205727, 210260, 210269, 211064, 215017, 215019, 215020, 215023, 215030, 215037, 215043, 215054, 215058, 215059, 215063, 215068, 215070, 215071, 215079, 216340, 216348, 216350, 216357, 216363, 216366, 216368, 216416, 216423, 216424, 216425, 219240, 219253, 219254, 219273, 219288, 219291, 219295, 232678, 232948, 232950, 232964, 232981, 232997, 232999, 233001, 233010, 233240, 233281, 233299, 233344, 233355, 233369, 233396, 233403, 233481, 233519, 233530, 233535, 233543, 233549, 233592, 233600, 233632, 233695, 233712, 233733, 233781, 233798, 234308, 238716, 238720, 271131, 271328, 271330, 271331, 276129, 276138, 276156, 276162, 276170, 276177, 276183, 276189, 276195, 276218, 276226, 276228, 276239, 276249, 276254, 276255, 276258, 276262, 276264, 276265, 279683, 282927, 282945, 283260, 285876, 285943, 285946, 285961, 285999, 286011, 286018, 286019, 286250, 295694, 295725, 297384, 297473, 297598, 297639, 299635, 299643, 299648, 299662, 299670, 299676, 299689, 300302, 300629, 300634, 300656, 300685, 300690, 300698, 300705, 300709, 300719, 300738, 300744, 300754, 300777, 300780, 300793, 300800, 300806, 300832, 300834, 300850, 300904, 300905, 300917, 300918, 300927, 00931, 300936, 300951, 300955, 300967, 301844, 301857, 301863, 301866, 301867, 301896, 301924, 301942, 301961, 301976, 301981, 302015, 302059, 302102, 302149, 302209, 302224, 302239, 302246, 302341, 304526, 305795, 306179, 307210, 312344, 312358, 312367, 312370, 312371, 312377, 312378, 312382, 312384, 312387, 312391, 312393, 312403, 312415, 312419, 312420, 312423, 312425, 312427, 312435, 312437, 312438, 312443, 312445, 312446, 312451, 313218, 317706, 317710, 317713, 317723, 317726, 317728, 317742, 317744, 317750, 317751, 317761, 317796, 317802, 317817, 317832, 318000, 318012, 318023, 318029, 318042, 318043, 318044, 318045, 318046, 318048, 318054, 318792, 321954, 321955, 321956, 321962, 321972, 321974, 321975, 321977, 321978, 321979, 321988, 321994, 322002, 322015, 322019, 322021, 322028, 322046, 322048, 322065, 322066, 322076, 322087, 322088, 322103, 322112, 322114, 322116, 322123, 322126, 322135, 322140, 322144, 322147, 322149, 322154, 322157, 322162, 322175, 322176, 322178, 322181, 322182, 322183, 322184, 322185, 322186, 322187, 322189, 322190, 322191, 322192, 322194, 322195, 322196, 322197, 322198, 322202, 322203, 322205, 322208, 322209, 322210, 322211, 322212, 322222, 322226, 322231, 322244, 322249, 322259,

322260, 322261, 322269, 322273, 322274, 322279, 322283, 322285, 322311, 323752, 323776, 323803, 323822, 323918, 323968, 324030, 324056, 324137, 324144, 324290, 324855, 324864, 325036, 325048, 325622, 325635, 325653, 325667, 325677, 325700, 325779, 325817, 325850, 325857, 325861, 325867, 325881, 325896, 325908, 325965, 326003, 326047, 326060, 326108, 326116, 326121, 326130, 326139, 326187, 326214, 330307, 330308, 330310, 330311, 330312, 330313, 330314, 330316, 330317, 330318, 330319, 330320, 330321, 330324, 330325, 330327, 330332, 330333, 330336, 330337, 330339, 330341, 330429, 330446, 330465, 330466, 330468, 330476, 330478, 330479, 330480, 330481, 330487, 330491, 330492, 330494, 330496, 330599, 330614, 330616, 330621, 330624, 330625, 330628, 330644, 331269, 331298, 333635, 338420, 338450, 338593, 338598, 338665, 384109 (384150 on tang), 386537, 386502, 386522, 86537, 386632, 386656, 396987, 398554, 398650 (386602 on tang), 398983, 401404, 401374, 21577, 423099.

Marlin Model 1894 or 94 carbine, cal. .44-40, with 15 inch barrel, S/Ns 375792, 386485, 398492, 398586, 407596, 421436, 425082, 427022.

Marlin, Model 1894 carbine, cal. .44-40, with factory fitted 15 inch barrel, S/Ns 386612, 425194.

Marlin Model 1894, caliber .44-40, S/N 376768, with 15.25-inch barrel.

Marlin Model 1895 or 95 carbine, cal. .38-56, with 15-inch barrel, S/N 167406.

Marlin Model 1895 or 95 carbine, cal. .40-65, with 15-inch barrel, S/Ns 313644, 315054.

Marlin Model 1895 or 95 carbine, cal. .45-70, with 15-inch barrel, S/Ns 313669, 313673, 313683, 313684, 313688.

Marlin Model 93, caliber .32 H.P.S., S/N 3375, with 15-inch barrel.

Mauser, Commercial Luger Artillery model semiautomatic pistols cal. 9mm, Mauser banner marked, produced under contract for the Royal Thai Police accompanied by original, German mfd., detachable wooden shoulder stocks.

Mauser, Model 1896 semiautomatic pistol accompanied by original German mfd. detachable wooden holster/shoulder stocks, all semiautomatic German mfd. variations produced prior to 1940, any caliber.

Mauser, Pistol-Carbine, model 1896, 7.63mm, with shoulder stock and 11-3/4" to 16" barrel.

Mauser, Model 1902, 6 and 10 shot magazine capacity, semiautomatic pistols in cal. 7.63 x 25mm (.30 Mauser), having the distinctive hammer safety, barrel lengths of either 3.9 or 5.5", and accompanied by an original detachable wooden holster/shoulder stock.

Mauser, Model 1912/14 pistol, 9mm Mauser short or .45 ACP, with original detachable wooden holster/shoulder stock and 5" barrel.

Mauser Parabellum 75th Year Commemorative Luger Karabiner with accompanying shoulder stock.

E. Mayer single shot handgun, no S/N, cal .38-55 smoothbore, barrel length 14-1/2".

Military type Nambu pistol, model 1904, cal. 8mm Nambu with an accompanying original detachable telescoping wooden holster/shoulder stock.

* Novelty four barrelled palm gun in .22 caliber, manufactured by O.F. Mossberg or C.S. Shattuck Arms Co, circa 1906-1909.

One Pocket Creedmores and other original pocket rifles with extension shoulder stocks, cal. .22, made by Samuel Watson Johnson (1838-1903).

OSS Glove Pistol, caliber .38 S&W or .38 special.

OSS "Liberator" pistol, .45 ACP or 9mm.

Remington Flare (Very) Pistol, Mark III, 10 gauge.

Royal, semiautomatic pistol, cal. 7.63 x 25mm (.30 Mauser) having an integral 10 or 20 round magazine, 5-1/2", 6-1/4", or 7-1/8" barrel, and accompanied by an original Spanish mfd. detachable holster/stock.

Sedgley, Mark V, 10 gauge, signal pistol.

The Shatuck "Unique" palm gun in .22 and .32 cal. rimfire.

Smith & Wesson, model Military and Police revolver, cal. .38, S/N 112037, with original Ideal holster/shoulder stock.

Smith & Wesson, model 40, Light Rifle, caliber 9mm parabellum.

Spanish Star, model A, semiautomatic pistol in cals. 7.63 Mauser, 9mm parabellum, 9mm long, .38 ACP and 45 ACP, accompanied by original Spanish mfd. detachable wooden holster/shoulder stock.

Spanish mfd. copies of the Mauser, model 1896, semiautomatic pistol produced prior to 1946 in cal. 7.63mm or 9mm and having either integral or detachable magazines; accompanied by

original Spanish mfd. detachable wooden holster/shoulder stock.
Stevens, Rifle, No. 20, w/smooth bore barrel for .22 and .32 rimfire shot cartridges.
Stevens, Reliable Pocket Rifle, second issue, cal. .22 long rifle or .22 Stevens-Pope.
Stevens, New model Pocket Rifle, first issue, caliber .22.
Stevens, New model Pocket Rifle, second issue, cal. .22 short, long or long rifle rimfire, .22 WRF, .32 long centerfire.
Stevens, New model Pocket Rifle No. 40, calibers: .22 long, rifle, .22 WRF, .22 Stevens-Pope, and .32 long centerfire.
Stevens, Hunter's Pet No. 34 Pocket Rifle, cal. .22 short rimfire to .44-40 WCF.
Stevens, Vernier Hunter's Pet No. 34 1/2 Pocket Rifle, cal. .22 short .44-40 WCF.
Stevens, Vernier New model Pocket Rifle 40 1/2, cal. .22 long rifle, .22 WRF, .22 Stevens-Pope, .32 long centerfire.
Stevens, 1898 New model Pocket Shotgun, in cals. .38-40 and .44-40.
Stevens, Number 39, New model Pocket Shotgun, cals. .38-40 and .44-40.
Stevens, 44-1/2 rifle, .22 cal. S/N 11006 with two-piece barrel owned by A.O. Neidner, Malden, MA.
Steyr Hahn, model 1911/12, semiautomatic pistol, cal. 9mm Steyr or 9mm parabellum with 5" barrel and accompanied by an original European detachable holsterstock.
Strong, Breech Loading Salute Cannon, 10 gauge, mfd. by Strong Firearms Co., New Haven, CT after 1898.
Swedish, model P-40 Lahti, semiautomatic pistol, in cal. 9mm parabellum, accompanied by original Swedish detachable wooden holster/shoulder stock.
The Taylor "Fur Getter" mfd. by the F.C. Taylor Fur Company, St. Louis, MO, .22 cal. rimfire.
The Taylor "Sure Shot" Trap Sun manufactured by or for the F.C. Taylor Fur Company, St. Louis, Missouri, .38 caliber and having a smooth bore.
U.S. Coast Guard 37mm Deck Cannon, S/N M1-1983.
U.S., Mark II, 10 gauge signal pistols.
Walther, model 1937, pistol, "Armee Pistols," 9mm parabellum, with original detachable shoulder stock and 4.9" barrel.
Webley & Scott pistol, Mark 1, No. 2, .455 cal., with original detachable shoulder stock.
Winchester Model 36, shotgun, 9mm rimfire.
Winchester rifle, Model 1873, .32 caliber, S/N 637480, having a 14-inch barrel.
Winchester Model 1873 carbine, cal. .32-20 WCF, with 14 inch barrel, S/N 695081 B.
Winchester Model 1873 carbine, caliber .32 WCF, with a 15-inch barrel, S/N 382028.
Winchester Model 1873 carbine, cal. .44-40 WCF, with 14 inch barrel, S/Ns 612999B, 710976B, 719510B, 88502.
Winchester Model 1873 carbine, cal. .44-40 WCF, with 15 inch barrel, S/Ns 579080B, ATF-33300-252.
* Winchester Model 1873 carbine, caliber .44 WCF, S/Ns 717912 B and 718977B, with 14-inch barrel.
Winchester Model 1885 carbine, cal. .25-35, with 15-inch barrel S/N 104783.
Winchester Model 1885 carbine, cal. .44-40 WCF, with 15 inch barrel, S/N 102429.
* Winchester Model 1885 rifle, cal. .44 WCF, S/Ns 75808 and 103509, having a 15-inch barrel.
Winchester Model 1885, caliber .44WCF, S/N 103509, with 15-inch barrel.
Winchester Model 1886 Line Throwing Gun, cal. .45-70 smoothbore, with 14-1/2 inch barrel, S/Ns 153830, 154228, 154229, 154231, 154232, 154256, 154257, 154258, 154259, 154260, 154262, 154263, 154264, 154265, 154266, 154267, 154268, 154269, 154270, 154271, 154274, 154287, 154301, 154304, 154305, 154308, 154319, 154320, 154321, 154322, 154327, 154328, 154331, 154332, 154334, 154336, 154337, 164338, 154339, 154340, 154342, 154345, 154346, 154347, 154348, 154349, 154350, 154351, 154352, 154355, 154356, 154357, 154359, 154360, 154365, 154366, 154384, 154385, 154388, 154390, 154393, 154396, 154397, 154398, 154402, 154403, 154404, 154406, 154409, 154411, 154413, 154414, 154415, 154418, 154419, 154420, 154422, 154424, 154427, 154428, 154429, 154437, 154438, 154439, 154440, 154450, 154451, 154452, 154453, 154455, 154463, 154464, 154465, 154466, 154467, 154468, 154471, 154476, 154511, 154512, 154514, 154515, 154516, 154517, 154520, 154521, 154522, 154523, 154524, 154525, 154526, 154527, 154528, 154529, 154531, 154532, 154548, 154550, 154551, 154552, 154554, 154555, 154556, 154557, 154558, 154562, 154564, 154565, 154566, 154567, 154568, 154569, 154570, 154572, 154573, 154574, 154588, 154589, 154590, 154591, 154592, 154595, 154598, 154599, 154600, 154601, 154602, 154604, 154605, 154607, 154608, 154609, 154610, 154611,

154612, 154613, 154616, 154617, 154618, 154619, 154627, 154629, 154631, 154633, 154639, 154640, 154642, 154646, 154648, 154650, 154651, 154653, 154654, 154655, 154660, 154661, 154663, 154664, 154665, 154666, 154669, 154671, 154672, 154676, 154677, 154682, 154688, 154689, 154690, 154693, 154694, 154696, 154697, 154698, 154700, 154702, 154704, 154707, 154708, 154709, 154712, 154713, 154714, 154715, 154718, 154719, 154720, 154721, 154722, 154724, 154725, 154726, 154730, 154731, 154732, 154735, 154737, 154739, 154741, 154742, 154744, 154745, 154747, 154749, 154755, 154764, 154768, 154769, 154771, 154773, 154775, 154776, 154777, 154778, 154779, 154780, 154781, 154782, 154789, 154790, 154791, 154793, 154794, 154795, 154797, 154798, 154799, 154800, 154801, 154802, 154803, 154804,154808, 154814, 154820, 154821, 154822, 154823, 154824, 154825, 154826, 154827, 154830, 154831, 154833, 154834, 154853, 154854, 154855, 154856, 154857, 154858, 154860, 154862, 154865, 154866, 154868, 154869, 154870, 154871, 154875, 154878, 154889, 154891, 154892, 154893, 154894, 154895, 154896, 154897, 154905, 154906, 154912, 154913, 154914, 154928, 154942.

Winchester Model 1892 or 92 carbine, cal. .25-20 WCF with 14-inch barrel, S/Ns 19782, 466396, 575644, 765952, 766291, 800432, *801205, 801232, 801389, 836858, 818777, 859905, 859912, 903891, 940986, 942997.

Winchester Model 1892 or 92 carbine, cal. .25-20 WCF with 15 inch barrel, S/Ns 295206, 512237, 692609, 846245, 852773, 853241, 859948, 982371, 850411.

Winchester Model 1892, caliber .30 WCF, S/N 211284, having a 14-inch barrel.

Winchester Model 1892 or 92 carbine, cal. .32 WCF with 14-inch barrel, S/Ns 353676, 605816, 800188, 807261, 840518, 874184, 928023, 939107.

Winchester Model 1892 or 92 carbine, cal. .32 WCF with 15-inch barrel, S/Ns *475522, 786820.

Winchester Model 1892 or 92 carbine, cal. .32-20 WCF with a 14-inch barrel, S/Ns 615287, 655804, 772706, 807261, 808290, 826022, 837181, 850460, 850556, 859920, 874501, 875174, 960471, 983151.

Winchester, Model 1892, caliber .32 WCF, S/N 841245, with 14-inch octagonal barrel.

Winchester Model 1892 or 92 carbine, cal. .32-20 WCF with a 15-inch barrel, S/Ns 203439, 472150, 475520, 490150, 554549, *690038, 800472, 807957.

Winchester, Model 1892, caliber .35 Winchester S.L., S/N 686515, with 15-inch barrel, marked .25-20 W.C.F.

Winchester Model 1892, cal. .38 WCF, with a 12-inch barrel, S/N 681490.

Winchester Model 1892 carbine, caliber .38 WCF, with a 14-inch barrel, S/Ns *426935, 675600, 686204, 697846, 842774, 848416, 848742, 998336.

Winchester Model 92 carbine, cal. .38 WCF, with a 14-inch barrel, S/Ns 672832, 979585.

Winchester Model 1892 carbine, caliber .38 WCF, with a 15-inch barrel, S/Ns 545168 and 600577.

Winchester Model 1892 or 92 carbine, cal. .38-40 WCF with 12 inch barrel, S/N 290849.

Winchester Model 1892 or 92 carbine, cal. .38-40 WCF with 14 inch barrel, S/Ns 288562, 475736, 653874, 667950, 679709, 847750, 848745, 848873, 850811, 887641, 975236, 998845.

Winchester Model 1892 or 92 carbine, cal. .38-40 WCF with 15 inch barrel, S/Ns 179252, 408307, 449677, 04878, 832039, 842788, 843000.

* Winchester Model 1892, cal. 44-40 WCF, S/N 242340, 999009 with 15-inch barrel.

Winchester Model 1892 carbine, cal. .44 WCF, S/Ns 499886, with factory fitted 14-inch octagonal barrel.

Winchester Model 1892, caliber .44 WCF, S/N 414816, with 14-inch octagonal barrel.

Winchester Model 1892 carbine, cal. .44 WCF, S/Ns *410040, 417473, 518749, with factory fitted 14-inch barrel

Winchester Model 1892 or 92 carbine, cal. .44 WCF, with 14-inch barrel, S/Ns 237383, *241998, 290545, 319311, 323486, 324906, 325125, 325258, 325412, 326732, 353128, 373701, 374202, 394598, 407490, 407735, *409734, 409959, 410015, 410127, 410461, 410488, 411197, 417104, 417274, 471250, 492863, 507295,558135, *567955, 600075, 613228, 659956, 661061, 676111, 688571, 760056, 710257, 742128, 742330, 760298, 761294, 767088, 814474, 845258, 852145, 853128, 861322, 870893, 879896, 880098, 882743, 889143, 894524, 895290, *895626, *896406, 940939,905438, 934212, 940138, *941030, 953433, 953533, 972059, 977206, 978960, 979529, 979915, 98094, 981253, 982178, 982703, 982509, 987936, 996120, ATF-33100012, 89287, 325676, 436742 and 982549.

* Winchester Model 1892, caliber .44 WCF, S/Ns *58732, 168291, 172068, 191413, 302989, 338220,

454616, 449861, 677916, 697497, 756846, 820410, 820507, 820970, 821110,824056, 827272, 835806, 842869, 844314, 850990, 871544, 882743, 889143, 893373, 894230, 895861, 908502, 971295, 985251, 991413, *992152, 995690, 996327, 997543, 998263, 998466, 998995, 998780, 999018, 999036, with 15-inch barrel.

* Winchester Model 1892, caliber .44, S/N 773086, having a 14-inch barrel with an engraved receiver.

Winchester carbine, Model 1892, caliber .44 WCF, barrel length 14 1/8 inches, S/N 895675.

* Winchester rifle, Model 1892, .44 WCF caliber, S/Ns having a 15-inch barrel.

* Winchester Model 1892, caliber .44 WCF, S/N 998929, with factory fitted 15-inch barrel.

Winchester Model 1892, caliber .44 WCF, S/Ns 408070, 693360 with 12-inch barrel.

Winchester Model 1892, caliber 44-40 WCF, S/Ns 172450, *301164, *678255, with 15-inch barrel.

Winchester Model 1892 or 92 carbine, cal. .44-40 WCF, with 14 inch barrel, S/Ns 268828, 319250, 320567, 323075, 323300, 323428, 324318, 325248, 325618, 326147, 340471, 349471, 407898, 409715, 410033, 417996, 446526, 475736, 495045, 583403, 600903, 710267, 743634, 761295, 819003, 820170, 822013, 823943, 824396, 842679, 852636, 861856, 879282, 890745, 895062, 895675, 897168, 898838, 898938, 921643, 925550, 948932, 976596, 977351, 977828, 978462, 979227, 979429, 980055, 980097, 980478, 981212, 982254, 982483, 982562, 984073, 987239, 987305, 989144, 993799, 999398.

Winchester Model 1892 or 92 carbine, cal. .44-40 WCF, with 15-inch barrel, S/Ns 197124, 381573, 414906, 511099, 518748, 650991, 659700, 756846, 816422, 820590, 823559, 837672, 844464, 881573, 886301, 895657, 896494, 905158, 937022, 954796, 977103, 988160, 990107, 991595, 994171, 996523, 997305, 998059, 998419, 998461, 998679, 998741, 998761, 998844, 998852.

Winchester Model 1892, caliber .25-20WCF, S/N 686515, with 15-inch barrel.

Winchester Model 1892 caliber .25-20WCF, S/N 914301, with 15-inch barrel.

Winchester Model 1894 or 94 carbine, cal. .25-35, with 14-inch barrel, S/Ns 427187, 814142, 865268.

Winchester Model 1894 or 94 carbine, cal. .25-35, with 15-inch barrel, S/Ns 155531, 444585, *522192, 542097, 589609, 812956, 815238.

Winchester Model 1894 or 94 carbine, cal. .30-30 WCF, with 14-inch barrel, S/Ns 462548, 505307, 793359, 805815, 842523, 864415.

* Winchester Model 94 lever action rifle, caliber .30 WCF, S/N 990233, with factory fitted 15-inch barrel.

Winchester Rifle, Model 1894, .30 WCF caliber, S/Ns 442920, 463348, 554548, 612946, 639534, 780670, 820363, 844538, 846610, 847112, *865212, 862579, 885127, 885359, 921311, 921754, 921922, *925038, 958505, 992289, 959632, 986535, 1004929, 1004427, 1005373, 1009473, 1017950, 1032419, 1044205, 1050280, 1054200, *1052693, 1056793, 1057984, 1058915, 1065744, 1086186, having a 15-inch barrel.

* Winchester Model 94 carbine, caliber .30 WCF, S/Ns 441082, 846992, 1009372, 1086226, 1009382, 865026, 870711 and 839968. with 15-inch barrel.

Winchester Model 1894 carbine, caliber .30 WCF, S/N 321470 with factory fitted 14-inch octagonal barrel.

* Winchester, Model 94 carbine, caliber .30 WCF, S/Ns 413187, 746707, 809300, 916487 with 14-inch barrel.

Winchester rifle, Model 1894, .30 WCF caliber, S/N 1057890, having a 15-inch barrel.

Winchester Model 1894 carbine, cal. .30 WCF, S/Ns *847116, *910606, 1017717, *1031654, 1033895, with factory fitted 15-inch barrel.

Winchester Model 94 carbine, cal. .30 WCF, S/N 662205, with factory fitted 15-inch barrel.

Winchester Model 1894 or 94 carbine, cal. .30 WCF, with a 15-inch barrel, S/Ns 502433, 503849, 505391, 507520, 639857, 640789, 758048, 801966, 820102, 864818, 885107, 885500, 909096, 910366, 923115, 923639, 937406, 959256, 981707, 992287, 998264, 1017931, 1021206, 1029281, *1033896, 1034950, *1038930, 1040702, *1041205, 1046882, 1040987, 1065446, 1065843, 1066548.

Winchester Model 1894, .30 WCF caliber, S/Ns 1052837, 933906, 910064, 922873 having a 15-inch barrel.

Winchester Model 1894 carbine, caliber .30 WCF, S/N 554734, with factory fitted 15 1/8-inch barrel.

Winchester Model 1894, cal. .30-30 WCF, S/N 848830, with original 15-inch unserviceable barrel.

Winchester Model 1894 or 94 carbine, cal. .30-30 WCF, with 15-inch barrel, S/Ns 273691,315250,

318257, 324318, 360587, 368971, 373033, 433426, 444269, 447421, 464604, 467286, 478529, 481877, 486201, 491869, 495342, 554439, 554727, 559831, 593839, 602691, 638701, 639264, 661329, 662192, 701730, 758406, 790018, 790761, 794467, 802743, 806780, 815570, 820101, 820375, 820413, 839880, 840123, 857252, 862245, 868769, 868912, 869576, 870637, 871116, 871276, 872505, 882967, 883055, 884272, 885127, 885975, 887409, 910758, 923215, 928280, 933642, 933902, 938370, 943237, 952507, 958812, 958947, 959421, 959486, 972583, 973001, 973063, 976472, 981452, 981456, 986621, 990851, 991114, 992557, 995540, 995578, 995675, 998829, 1004615, 1004958, 1009373, 1009414, 1013465, 1025302, 1032798, 1033898, 1033973, 1034037, 1040905, 1046737, 1051727, 1051776, 1052211, 1052749, 1052851, 1052982, 1057043, 1059318, 1066951, 1068292, 1072755, 1080781, 1086226, 1090460

Winchester Model 1894 or 94 carbine, cal. .32 Win. Spl., with 14-inch barrel, S/N 781211, 889043.

Winchester Model 1894 or 94 carbine, cal. .32 Win. Spl., with 15-inch barrel, S/Ns 437440, 840832, 859421, 860098, 880845, 1006715.

Winchester Model 1894 or 94 carbine, cal. .32-40 WCF, with 15-inch barrel, S/N 355640, 894588.

Winchester Model 1894 or 94 carbine, cal. .38-55 WCF, with 15-inch barrel, S/N 247646.

Winchester, 1902, pistol made from a rifle, cal. .22, produced at factory, having no S/N, but having a decorated stock and accompanied by documents denoting its use by Mr. Ned Tupperwein.

Winchester, 1892 Saddle Ring Carbine, .32 WCF cal. with 15" barrel, S/Ns 575423 and 838742.

Winchester, 1892 Saddle Ring Carbine, .38 WCF cal. with 14" barrel, S/N 683395.

Winchester, 1892 Saddle Ring Carbine, .44 WCF cal. with 14" barrel S/N 988293.

Winchester, 1894 Saddle Ring Carbine, .30 WCF cal. with 15" barrel S/N 959111, 1040690 and 986607.

Winchester, Model 1894 Saddle Ring, .30 WCF cal. with 15 1/16 " barrel S/N 933879.

SECTION IIIA: Weapons Removed From The NFA As Collector's Items And Removed From the GCA As Antiques.

The following firearms were removed from the NFA as collector's items and classified as curios or relics under 18 U.S.C. Chapter 44. However, because they are antiques as defined in Chapter 44, they should not have been classified as curios or relics. Because they are no longer NFA weapons and are antiques under Chapter 44, they are not subject to GCA provisions.

Any pistol or revolver, mfd. in or before 1898, originally designed to accept a shoulder stock, and accompanied by an original shoulder stock.
Belgian, Cane Gun, 41 caliber rimfire.
Bergmann, model 1897, cal. 7.65mm (7.8mm) pistol with accompanying shoulder stock.
Borchardt, model 1893, cal. 7.63mm pistol with accompanying shoulder stock.
Chicago, palm pistol, caliber .32 rimfire extra short.
Frank Wesson, Bicycle Rifle with accompanying shoulder stock.
French, Model 1897 75mm breechloading cannon of French manufacture, mfd. in or before 1898.
Gaulois, palm squeezer, 8mm short "Little All Right" palm pistol, .22 cal. rimfire patented by Edward Boardman and Andrew Peavey, January 18, 1876.
German Minenwerfers, old and new models, mfd. prior to 1919, utilizing radial, straight pull friction primer ignition system.
Mannlicher, Pistol-Carbine, model 1896, 7.63mm Mannlicher, with rifle type shoulder stock and forearm and 11-3/4" barrel.
Marlin Model 1889, cal. .32, with 15-inch barrel, S/Ns 45952, 69571, 70624, 70678, 70729, 70737, 70738, 80884, 85429, 96169.
Marlin Model 1889, cal. .38, with 15-inch barrel, S/Ns 53545, 63448, 63476, 63478, 69566, 69568, 69569, 69657, 69660, 73133, 73715, 73742, 83805, 83875, 83892, 83949, 83957, 83984, 83986, 88942.
Marlin Model 1889, cal. .44, with 15-inch barrel, S/Ns 40013, 40021, 40074, 40083, 40084, 40089, 44305, 44306, 44313, 44324, 44327, 44330, 44332, 44335, 44336, 44338, 44343, 44345, 44347, 44348, 44351, 44360, 44366, 44369, 44372, 46621, 46641, 46709, 46743, 46765, 46876, 48997, 49009, 49012, 49016, 49019, 49026, 49029, 49031, 51902, 51912, 52022, 52023, 52029, 52031,52038, 52039, 52043, 53304, 53318, 53360, 53376, 53392, 53393, 55365, 55820, 59140, 59792, 61252, 61265, 61347, 61350, 61903, 61904, 61929, 61932, 61935, 61972, 61982, 61985, 63440, 63458, 63732, 63735, 63747, 63761, 63763, 65070, 65074, 66329, 66415, 66819, 66841, 66842, 66846, 66852, 66888, 67397, 67402, 67403, 67406, 67433, 67436, 67454, 67457, 67460, 67513, 67518, 68599, 68612, 68620, 70779, 70784, 70821, 70830, 70832, 72848, 73049, 73083, 73085, 73090, 73151, 73159, 73160, 73181, 73195, 73208, 73216, 73218, 73713, 73729, 73733, 73735, 73749, 73750, 73751, 73764, 73768, 73772, 73774, 73776, 73789, 73814, 73824, 73838, 73842, 73848, 73853, 81046, 81052, 81063, 81077, 81084, 81092, 81099, 81119, 81187, 81194, 83456, 83471, 83472, 83484, 83493, 83499, 83501, 83513, 83514, 83515, 83520, 83533, 83542, 83543, 83554, 83559, 83632, 83679, 83691, 83859, 83870, 83930, 83933, 83937, 83950, 83953, 83954, 83964, 83969, 86458, 86506, 86674, 86697, 86745, 86746, 86782, 86786, 86795, 88938, 89484, 89485, 89491, 89503, 89523, 89534, 89535, 89541, 89562, 89584, 89587, 89609, 89818, 89823, 89831, 90012, 90013, 90016, 90021, 90037, 90053, 90055, 90072, 90075, 90706, 91389, 91496, 91520, 91657, 91672, 91819, 91842, 91869, 91887, 91889, 91918, 91949, 91962, 91974, 91981, 91983, 91984, 91990, 91995, 92009, 92010, 92011, 92012, 92030, 92044, 92061, 92074, 92078, 92079, 92104, 92106, 92116, 92133, 92137, 92142, 92144, 92154, 92170, 92172, 92178, 92193, 92197, 92198, 92213, 92218, 92224, 99751, 99753, 99758, 99766, 99819, 99820, 99869, 99885, 99967, 99972, 99974, 99992, 99994, 100003, 100009, 100021, 100086, 100228, 108487, 108513, 108517, 108526, 108532, 108537, 108541, 108546, 108554, 108555, 108557, 108567, 108570, 108571, 108572, 108575, 108577, 108579, 108581, 108585.
Marlin Model 1893, cal. .25-36, with 15-inch barrel, S/N 124940, 374465.
Marlin Model 1893, cal. .30, with 15-inch barrel, S/Ns 161903, 162931, 164888.
Marlin Model 1893, cal. .32, with 15-inch barrel, S/Ns 96935, 103641, 103643, 107944, 107948, 107950, 107958, 107970, 124405, 131030, 142050, 147644, 147669, 162785, 162810, 176191.
Marlin Model 1893, cal. .38-55 or .38-56, with 15-inch barrel, S/Ns 96930, 96938, 96942, 103642, 103648, 103660, 103661, 103666, 103670, 133880, 147638, 147658, 167090, 167406.
Marlin Model 1894, cal. .25, with 15-inch barrel, S/Ns 137247, 137248, 137250, 137252, 137256, 137261, 140941, 146754, 146755, 146756, 146757, 146758, 146759, 147678, 147681, 147682, 147688, 147694,

147699, 147700, 147701, 147703, 148157, 148173, 148183, 148217, 151981, 174745.

Marlin Model 1894, cal. .32, with 15-inch barrel, S/Ns 47431, 84993, 126545, 129211, 129221, 129227, 130657, 134113, 134118, 134126, 135406, 135407, 137124, 137125, 137129, 137130, 137131, 137133, 137136, 137137, 137138, 137139, 137145, 137146, 137147, 137149, 137151, 137156, 140876, 140877, 140892, 140899, 140903, 140907, 140909, 140921, 140922, 140924, 140927, 140931, 140934, 140942, 140943, 140945, 142038, 142039, 147206, 147704, 148147, 148153, 148155, 148158, 148178, 148183, 148184, 148194, 148197, 148200, 148201, 148207, 148209, 148212, 148214.

Marlin Model 1894, cal. .38 WCF, with a 15-inch barrel, S/N 150463.

Marlin Model 1894, cal. .38, with 15-inch barrel, S/Ns 109874, 111674, 115770, 115772, 126799, 126858, 126907, 126953, 127093, 127180, 127382, 133190, 133260, 133491, 133645, 134348, 134363, 134364, 134446, 134495, 134526, 134565, 137669, 142371, 142374, 142379, 142389, 145783, 145784, 145787, 145788, 145789, 145790, 145792, 145793, 145794, 145796, 145798, 145799, 145800, 145806, 145813, 147629, 147630, 147631, 147632, 147633, 147634, 150482, 150483, 150486, 150576, 151586, 151594, 151651, 151721, 151726, 151737, 151746, 151748, 151751, 151753, 151761, 151923, 151987, 152004, 152043, 152057, 152066, 152070, 169850,169851, 169852, 169855, 169856.

Marlin Model 1894, cal. .44, with 15 inch barrel, S/Ns 109588, 111398, 111696, 111789, 111794, 111813, 115362, 115363, 115365, 115378, 115383, 115385, 115388, 115392, 115393, 115400, 115403, 115407, 115408, 115412, 115413, 115414, 115418, 115419, 115421, 115433, 115436, 115438, 115439, 115449, 115451, 115453, 115457, 115458, 115459, 115473, 115476, 116487, 115489, 115496, 115499, 115616, 115622, 115644, 115649, 115651, 115662, 115710, 115723, 115743, 115822, 120065, 120094, 120107, 120115, 120118, 120119, 120126, 120133, 120137, 120139, 120145, 120146, 120149, 120161, 120213, 120227, 120228, 120239, 120243, 120270, 120273, 120276, 120292, 120296, 120469, 120496, 120504, 120505, 120523, 120534, 120549, 121549, 121933, 124651, 124662, 124671, 124704, 124725, 124734, 126787, 126843, 126851, 127149, 127251, 127300, 133605, 134090, 134376, 134543, 137495, 137614, 137725, 137734, 137765, 142367, 142368, 142373, 142376, 142377, 142380, 142390, 142402, 144194, 144197, 144198, 144272, 144296, 145687, 145689, 145693, 148154, 150181, 150260, 150369, 150371, 150380, 150384, 150385, 150406, 150420, 150421, 150430, 150462, 150471, 150568, 151047, 151074, 151082, 151086, 151087, 151091, 151092, 151094, 151096, 151097, 151101, 151107, 151133, 151134, 151137, 151140, 151144, 151145, 151151, 151158, 151187, 151193, 151195, 151214, 151227, 151242, 151243, 151619, 151622, 151631, 161632, 151698, 151723, 151724, 151730, 151733, 151735, 151743, 151744, 151756, 151757, 151760, 151764, 151790, 151792, 151797, 151994, 151997, 152020, 152022, 152075, 169854, 169914, 169940, 169971, 169972, 169986, 170002, 170025, 170138.

* Mauser, Model 1896, semiautomatic pistols or carbines having the following combination of features: 1.) Serial number below 1500. 2.) Large ring or cone hammer. 3.) Early safety (safety on with lever in down position). 4.) Early extractor (straight sided at front without semicircular lobes on either side). 5.) Early firing pin retainer (U-shaped insert in back of bolt which retains firing pin).

Mervellieux, squeezer pistol, 6mm and 8mm short.

Peavey, A.J., Knife Gun, .22 short rimfire.

Protector palm gun, .32 rimfire extra short, patented by Jacques Tublaux, Patent No. 732644.

Quackenbrush Bicycle Rifle, with telescopic wire stock, .22 caliber.

Remington, Cane Gun, model 1, .22 rimfire.

Remington, Cane Gun, model 2, .32 rimfire.

Springfield Trapdoor, Model 1884, cal. 45-70, line throwing gun, having a barrel length less than 16 inches, S/N 294879.

Stevens, Old model Pocket Rifle, .22 short or long rimfire.

Stevens, Reliable Pocket Rifle, first issue, cal. .22 short, long or long rifle.

Stevens, New Model Pocket Rifle, first issue, in cal. .32 short or long rimfire.

Stevens, New Model Pocket Rifle, 2nd issue, in cals. .25 Stevens or .32 long rimfire.

Stevens, New Model Pocket Rifle No. 40, in cal. .25 Stevens or .32 long rimfire.

Stevens, New Model Pocket shotgun, in cal. .44-50 Everlasting.

Stevens, Vernier, New Model Pocket Rifle, cal. .22 short, .22 long rifle, .22 WRF, .32 long rimfire.

Stevens, Vernier, New Model Pocket Rifle, No. 40 1/2, in cal. .25 Stevens or .32 long rimfire.

Tribuzio, "Squeezer" invented by Catello Tribuzio of Turin, Italy, cal. 8mm short.

Winchester Model 1873 carbine, cal. .32-20 WCF, with 15-inch barrel, S/N 382027.

Winchester Model 1873 carbine, cal. .38-40 WCF, with 14-inch barrel, S/N 514709.

* Winchester Model 1873, caliber .44 WCF, S/Ns 131592A and 92835A, with 15-inch barrel.

* Winchester Model 1873 carbine, cal. .44-40 WCF, with 15-inch barrel, S/Ns 69450, 92537, 92842A, 127884,

131762A, 145337A, 198040B, 234795B, 235014B, 247094B, 263826B, 336514B, 336549, 380061, 486139B.
Winchester Model 1885 carbine, cal. .38-40 WCF, with 15-inch barrel, S/N 79085.
Winchester Model 1885 carbine, cal. .44-40 WCF, with 15-inch barrel, S/Ns 73604, 73885, 78928, 79256, 83304.
* Winchester Model 1885 single shot carbine, cal. .44 WCF, with factory fitted 15-inch barrel, S/N 83313.
Winchester Model 1885 carbine, caliber .44 WCF, S/Ns 73593, 75744, *75759, with a 15-inch barrel.
Winchester Model 1892 carbine, cal. .32-20 WCF, with 14-inch barrel, S/N 50480.
Winchester Model 1873 Carbine, caliber .44 WCF, having a 15-inch barrel, S/N 249401B.
Winchester Model 1892 carbine, cal. .38-40 WCF, with 15-inch barrel, S/N 43844.
Winchester Model 1892, caliber .38 WCF, S/N 102645, with 15-inch barrel.
Winchester Model 1892 carbine, cal. .44-40 WCF, with 15-inch barrel, S/Ns 51976, 95516, 158145.
Winchester Model 1894 carbine, cal. .30-30 WCF, S/N 42771.

SECTION IV: National Firearms Act Weapons Classified As Curios Or Relics Under 18 U.S.C. Chapter 44

The Bureau has determined that the following NFA weapons are curios or relics as defined in 27 CFR 478.11 because of their dates of manufacture. These NFA weapons, classified as curios or relics, are still subject to all the controls under the NFA. However, licensed collectors may acquire, hold, or dispose of them as curios or relics subject to the provisions of 18 U.S.C. Chapter 44 and 27 CFR Part 478. They are still "firearms" as defined in 18 U.S.C. Chapter 44 and 26 U.S.C. Chapter 53.

Alarm Pistol, S/N IRS-3591, cal. .22 rimfire, marked Patent Pending about 1883, which rings an alarm bell and also fires a cartridge when set to do so.
Armalite, AR 10, 1955, cal. 7.62mm NATO submachinegun, made by the Armalite Div. Of Fairchild Engine & Airplane Co., Costa Mesa, CA, prior to 1961.
Armalite AR-15, .222 Remington or.223 cals., produced by Armalite.
Armalite AR-15, Model 601,.223 cal., mfd. by Colt. Must be marked "Armalite."
Armalite, AR-18, selective fire rifle, cal. .223, prod. by Armalite, Costa Mesa, CA.
Astra, models 901, 902, and 903, selective fire pistols, cal. 7.63mm (.30 Mauser).
Astra, model F, selective fire pistol, caliber 9mm Bergmann.
Azul, selective fire pistol, cal. 7.63mm (.30 Mauser), copy of Mauser model 1896 pistol.
Australian, Austen MK I and MK II Submachineguns, cal. 9mm, mfd. in Australia prior to 1946.
Australian, Owen MK I and MK II Submachineguns, cal. 9mm, mfd. in Australia prior to 1946.
Austrian, Swartzlose machinegun, model 1905 and 1907/1912, cal. 8mm.
Austrian, Steyr-Solothurn model MP-30 and MP-34 submachineguns as originally mfd. by Steyr-Daimier-Puch of Steyr-Austria, prior to 1946.
Auto Ordnance, WWII commemorative Thompson model 1928, caliber .45.
Auto Ordnance, Korean War Commemorative Thompson model 1928, cal. .45.
Auto Ordnance, Thompson Model 1928, .22 caliber machinegun in original configuration as manufactured at the West Hurley, NY, facility between 1980 and 1981.
Benet Mercie, machine rifle, model 1909 (French and American manufacture), caliber 8mm Lebel or .30.
Boys, MK1, caliber .55 , anti-tank rifle.
Bren, light machineguns, any model, any cal., British Commonwealth manufacture.
Brevettata, model F.N.A.- B 1943, submachinegun, cal. 9mm parabellum,with an 8" barrel, S/N 5391.
British, Lanchester (Lancaster) Mk 1 and Mk 1, 9mm, submachineguns mfd. in England during World War II.
British Model 1903, 4.5 inch howitzer.
British, STEN submachineguns Mk. I, Mk. II, Mk. III, Mk. IV and Mk. V, cal. 9mm, original British Commonwealth Military issue only.
Browning automatic rifle (BAR), model 1918A2, mfd. by New England Arms Corporation prior to 1946.
Browning automatic rifle M1918A2, S/N 00229, mfd. by Group Industries.
Canadian, EX1 and EX2 experimental series of the FN FAL in cal. 7.62mm NATO, as mfd. by Fabrique Nationale, Herstal, Belgium, for the Canadian Government.
Cane Gun, cane with horn handle, silver tipped, wood-covered steel, breech loading, smooth bore barrel, cal. .32 centerfire, S/N IRS-5834.
Cane Gun, English walking stick, bamboo covered steel rifled barrel with horn handle, caliber .38 centerfire, S/N IRS-3589.
Cane Gun, cal. .410 centerfire, S/N IRS-3587.
Chinese mfd. copy of the U.S. 60mm mortar mfd. prior to 1946.
Chinese, Type 24 (Maxim) machinegun, cal. 7.92mm, Chinese manufacture.
Chinese, Type 26 light machinegun.
Colt, CMG-2, S/N 012.
Colt, Machinegun, model 1895, any caliber.
Colt, Machinegun, model 1914, any caliber.
Colt, Model MG38 and MG38B, water cooled machinegun any cal. mfd. by Colt.
Colt, Model 1919 auto rifle.
Colt, Model 1921, Model 1921A1, Model 52, and Model 52A, .50 cal., water-cooled machinegun, as originally mfd. by Colt and bearing the original S/N with the letter "C" prefix.
Colt, Monitor model machine rifle, cal. 30-06, commercial version of the BAR, 18" barrel, vertical pistol

grip, Cutts compensator, mfd. by Colt.
Crescent, Certified Shotgun .410 gauge, with 12-1/4" barrel and pistol grip.
Czechoslovakian, model ZB 26 and ZB 30 light machinegun.
Danish HOVEA M1949, 9mm parabellum, produced in Denmark.
Danish, Madsen machineguns, all models produced prior to 1946; all calibers.
Danish, Madsen, model 1946, submachineguns.
Danish, Madsen, model 1950, submachineguns.
Degtyarev RPD Light Machine Gun, 7.62 x 39mm caliber, mfd. in the Soviet Union, the People's Republic of China, or the People's Republic of Korea (North Korea).
Degtyarev-PTRD 1941, 14.5mm anti-tank rifle.
Egyptian "Port Said" version of the Swedish Model 45B, 9mm cal., produced in Egypt.
Enfield, military bolt action rifle cut-a-ways, having barrels of less than 16".
Fabrique Nationale, model BAR, full automatic rifle, all cals., as mfd. by Fabrique Nationale, Herstal, Belgium.
Federal Laboratories, 12 Gauge Gas Billy, all variations.
French, 25mm anti-tank cannon SA-L mil. 1937, commonly known as French models 34 SA and 37A.
French, Chatellerault model 1924/29 machineguns, cal. 7.5mm.
French, C.S.R.G. (Chauchat) machinegun model 1915, cal. 8mm Lebel.
French, C.S.R.G. (Chauchat) machinegun, model 1918, cal. .30.
French, model MAS 1938, sugmachinegun, caliber 7.65mm Long.
French, model MAT 49, submachinegun, caliber 9mm.
French, model 1897 75mm breechloading cannon of French and American manufacture, in 1899 or later.
French, model 37, 5Omm (1.97 in.) mortar - also designated as a German, model 203(f), 5cm (1.97 in.) mortar.
French, model 34 SA and 37A, antitank cannon 1937 in 25mm.
French, model 1934 M39 (Mle 34 K39), cal. 7.5 x 54mm, aircraft machinegun.
Franchi, model 1962 Semiautomatic carbines, caliber 9mm.
French, St. Etienne machinegun, model 1907, caliber 8mm. Lebel.
Finnish, Tampelia (original), model 1938, caliber 81mm mortars.
German, assault rifles selective fire, produced during 1941 through 1945, in caliber 7.92mm Kurz (7.92 x 33mm).
German, Bergmann model 1915 water cooled machinegun, caliber 7.92mm.
German, Bergmann model 1915, an air cooled machinegun, caliber 7.92mm.
German, Bergmann, MP-34, 9mm, machine pistol.
German, Bergmann, MP-35/1, 9mm, machine pistol.
German, BUCO hand firing device, caliber 12mm (.44 cal.).
German, Dreyse model 1918 water cooled machinegun, cal. 7.92mm.
German, Erma (EMP), or MPE, 9mm, machine pistol.
German Flak 30, 2cm anti aircraft machinegun.
German, Gast machinegun, caliber 7.9mm.
German, FG42 Parachutists rifle, cal. 7.92mm, mfd. prior to 1946.
German, MG 15 anti aircraft machinegun 7.92mm made before 1945.
German, MG 15 machineguns, cal. 7.92mm, made prior to 1945.
German mfd., Model MP3008 submachinegun, cal. 9mm Luger, produced in Germany c. 1944-1945.
German MP41 submachineguns, cal. 9mm parabellum, mfd. by C.G. Haenel, Suhl, Germany during WWII. (Original WWII German production only.)
German Minenwerfers, having fixed firing pins, spring or hammer actuated striker, any cal., any model, mfd. prior to 1919.
German, 15cm Nebelwerfer, 41 (150-mm) Rocket Launcher.
German, Parabellum machinegun, model 1913, 1914 and LMG 14/17, caliber 7.9mm.
German, Raketenpanzerbuchse 43 (8.8cm, RPzB43), model 43, 88mm Rocket Launcher.
German, Raketenpanzerbuchse 54 (8.8cm, RPzB54), model 54, 88mm Rocket Launcher.
German, Schmeisser, MP-28/II, 9mm machine pistol.
German, Steyr-Solothurn, 9mm machine pistol.
German, MG 34 light machinegun, cal. 7.92mm.
German, MG 42 light machinegun, produced prior to 1946, cal. 7.92mm.
German, MG 131 machineguns, all models, all calibers, mfd. before 1946.
German, MP 18-1, submachinegun, cal. 9mm parabellum.

SECTION IV

German, MP 38 submachinegun, cal. 9mm parabellum, mfd. by B. Geipel (Erma) prior to 1941.
German, MP 40 submachineguns, cal. 9mm, mfd. from 1940-1945 for the German military forces.
German, 7.5cm Pak 40 anti-tank gun (75mm).
German, 7.5cm Pak 39 anti-tank gun (75mm).
German, 7.5cm Stuk assault gun (75mm).
German, 7.5cm Kw.K 40 tank gun (75mm).
German, 8mm MG 81 and MG 81Z aircraft Machineguns.
German, 15/20mm MG 151 Aircraft Machineguns (all variations).
German, 30mm MG 108 Aircraft Machineguns/Cannons.
German, 13mm Rheinetal-Borsig model 131, Aircraft Machineguns (all variations).
German 37 mm Rheinmetall Borsig anti tank guns mfd. prior to 1946, including the following variations or designations: German issue 3.7cm PAK 35, PAK 35/36, PAK 36, Swedish 3.7cm PvKan m/38, Italian 37mm Cannone contracarro da 37/45 [captured designation 3.7cm PAK 162(I)], Russian 37mm Protivotankovaya Pushka obrets 1930 [captured desgination 3.7 PAK 158(r)], Dutch 37mm Rheinmetall [captured designation 3.7 cm PAK 153(h)], and Japanese 37mm Type 97 (PAK36 purchassed by Japan).
H&A A. Company, 16 gauge, set gun.
H&R Reising Model 55, submachineguns, caliber .45 ACP, manufactured by Harrington and Richardson, Worcester, MA, prior to 1946.
Harrington and Richardson (H&R) Handy Guns, smoothbore pistols, mfd. at Worcester, MA, all calibers, all barrel lengths, having an overall length of less than 26 inches.
Harrington and Richardson (H&R) Handy guns, with shoulder stock, mfd. at Worcester, MA, having a smoothbore barrel less than 18 inches in length, or having a rifled barrel less than 16 inches in length.
High Standard, USA model HD, .22 1r cal. pistols, originally equipped with silencers for issue to the OSS and other military agencies, S/N range 109110-153890.
Hotchkiss Field Cannon, Model 1909, 37mm caliber, S/N IRS 3857.
Hotchkiss, light machinegun, British model 1909 Mark I and Mark I, caliber .303.
Hotchkiss, models 1897, 1900 and 1914 machineguns of French mfd. in all calibers.
Ingram Models 6 and 7, various cals., produced by Police Ordnance Company.
Ingram Models 8 and 9, various cals., produced in Thailand.
Italian, Breda model 1930 machineguns, all calibers.
Italian, Breda Safat, model 1935, 7mm aircraft machinegun, mfd. by Societa Anonima Ernesto Breda, Brescia, Italy, prior to 1946.
Italian, Breda model 1937 heavy machinegun, caliber 8mm.
Italian, Cannone anticarro e d'accompagnamento 47/32 M1935, cal. 47mm, all variations and manufacturers.
Italian, Fiat Revelli model 1935 machinegun, caliber 8mm.
Italian, Model TZ-45, cal. 9mm submachineguns mfd. before 1946.
Italian, model 38A, 9mm parabellum submachinegun, mfd. by Pietro Beretta SpA, Gardone-val Trompia, Italy, prior to 1946.
Italian, model 38/42, 9mm parabellum, by Beretta SpA, Gardone-val Trompia, Italy prior to 1946.
Italian, Revelli model 1914 machinegun, caliber 6.5mm.
Italian, Scotti machineguns, all models, all calibers, mfd. Before 1946.
Italian, Villar Perosa submachinegun, model 1915, 1916 and 1917, cal. 9mm parabellum.
Ithaca, "Auto and Burglar Guns," mfd. by Ithaca Gun Company from 1922, to 1933, all original guns.
Ithaca Auto & Burglar Gun, 28 gauge, S/N 354442, mfd. in 1921 or 1922.
Japanese, (model 1921 Browning Type), aircraft fixed machinegun, cal. 12.7mm.
Japanese, experimental submachinegun "Bullpup" caliber 8mm Nambu.
Japanese, model 10 and 89, 50 mm grenade dischargers.
Japanese, model 38 Field Cannon.
Japanese, model 94, 37mm gun mfd. prior to 1946.
Japanese, model 94, 37mm gun, mfd. prior to 1946.
Japanese, modified Bergmann (SIG), submachinegun, cal. 7.63mm, featuring bayonet mounting lug.
Japanese, M1932 76mm Battalion Pack Howitzer.
Japanese, Type 1 (1941), aircraft flexible machinegun, cal. 7.92mm, twin barrel type.
Japanese, Type 3 (1927), heavy machinegun, cal. 6.5mm. twin barrel type.
Japanese, Type 11 (1922) light machinegun, all variations, caliber 6.5mm.
Japanese, Type 38 (105) machinegun (Hotchkiss) cal. 6.5mm.

Japanese, Type 89 (1929) Vickers pattern machineguns, cal. 7.7mm.
Japanese, Type 91 (1931), tank machineguns (modified Type 11), cal. 6.5mm.
Japanese, Type 92 (1932), heavy machinegun, Hotchkiss pattern cal. 7.7mm.
Japanese, Type 92 (1932), light machinegun or aircraft machinegun, Lewis pattern, cal. 7.7mm.
Japanese, Type 92 (1932), tank machinegun, caliber 13.2mm.
Japanese, Type 93 (1938), aircraft machinegun cal. 7.29mm.
Japanese, Type 94, 37mm gun, mfd. prior to 1946.
Japanese, Type 96 (1936), light machinegun cal. 6.5mm.
Japanese, Type 97 (1937), tank machinegun, cal. 7.7mm.
Japanese, Type 97, 20mm Anti-Tank machinegun.
Japanese, Type 98, 7.92 aircraft machinegun, Japanese mfd. prior to1946.
Japanese, Type 99 (1939), light machinegun, all varieties, cal. 7.7mm.
Japanese, Type 100 (1940), aircraft flexible machinegun, cal. 7.92mm, twin barrel type.
Japanese, Type 100 submachinegun, caliber 8mm Nambu, all varieties.
Japanese, 12.7 (13) mm Browning HO-103 machineguns.
Japanese, 20mm Browning Aircraft HO-5 machineguns/cannons.
Japanese, 20mm Oerlikon model 99, machineguns/cannons.
Japanese, 30mm Browning Aircraft HO-155 Type 1, machineguns/cannons.
Japanese, 30mm Browning Aircraft HO-155 Type 2, machineguns/cannons.
Japanese, 37mm Browning Aircraft HO-204 machineguns/cannons.
Japanese, 40mm Caseless Aircraft machineguns/cannons.
Lahti, model 39, caliber 20mm anti-tank rifle.
Lewis, light machinegun (American or British manufacture), caliber .303 or .30.
Lovelace Fur Getter; patented by C.D. Lovelace, June 2, 1914, mfd. by Hopkins and Allen Company, Norwich, CT, 12 gauge.
Luger, Mauser commercial manufacture, semiautomatic pistol, 75 Jahre, Parabellum-Pistol, Mod. Karabiner, Commemorative, caliber 9mm, accompanied by a carbine type shoulder stock.
Madsen light machineguns, Series A, all calibers, mfd. in Denmark between 1936 and 1956.
Marble's Game Getter, Firearms, with a combination of .44 cal. or .410 gauge smooth bore and .22 cal. barrels, both lst and 2nd models, with barrel lengths of less than 18", mfd. by the Marble Arms Corporation of Gladstone, MI, prior to 1943.
Marlin, aircraft machinegun, model 1917 and 1918, caliber .30.
Marlin, U.S. model 1917, ground machinegun, caliber 30/06.
Marlin, tank machineguns, model 1918 caliber .30.
Mauser, mfd. Schnellfeuer selective fire pistols and all original prototypes, cal. 7.63mm Mauser.
Maxim, machineguns of German manufacture, all models and cals.
Maxim, machineguns produced by Maxim Nordenfeldt and Vickers Sons and Maxim from approximately 1885 through 1908, all calibers.
Maxim, machineguns, Russian model 1905, Russian manufacture having a brass water jacket in caliber 7.62mm, Russian rimmed.
Oerlikon, MG-FF and MG-FFM, 20mm aircraft cannon.
Oerlikon, 20mm Automatic Cannon, all variations mfd. in the United States prior to 1946.
* Phoenix Cartridge Company, Model PCMP submachinegun, caliber 9mm or .45, S/N 2.
Pipe Pistol, S/N 420, cal. .22 rimfire (word "pipe" means smoking tobacco type).
Pipe Pistol, S/N IRS-3579, cal. 7mm centerfire, marked drgm Bucksom (word "pipe" means smoking tobacco type).
Reinmetall MG17, 7.9mm aircraft machinegun.
Reising, submachineguns U.S. model M-50, .45 cal., Harrington & Richardson Arms, Worchester, MA, USA, Pat Pending.
Rexer, original British made Madsen type light machineguns mfd. at the British Arms Factory known as Rexer.
Royal, selective fire pistol, cal. 7.63mm (.30 Mauser), copy of Mauser model 1896 pistol.
Russian, Degtyarev machineguns, models DA, DP, DPM, DT, and RP46, caliber 7.62mm.
Russian, model 1910 (Maxim), machinegun, any cal., Russian manufacture.
Russian, model PPD 1943/38 and PPD 1940, submachineguns, cal. 7.62mm, original Russian manufacture.
Russian, 50mm Mortar Launcher.
Russian, PSH41, 7.62mm submachineguns and all copies/variations produced in the Communist Bloc countries prior to mid 1950's.

Russian, PPS42, submachineguns, caliber 7.62mm.
Russian, PPS43 submachineguns, cal. 7.62mm mfd. in the Soviet Union prior to 1946.
Russian, model SG43 heavy machinegun, cal. 7.62mm, Russian manufacture.Russian (U.S.S.R.), Federov, automatic rifle, model 1916 AVF (Avtomat), cal. 6.5 x 5O.5SR (6.5mm Japanese), of Russian manufacture.
Russian (U.S.S.R.), Simonov, automatic rifle, model 1936 AVS (AVS36), cal. 7.62 x 64R, of Soviet manufacture.
Russian (U.S.S.R.), Tokarev, automatic rifle, model 1940 AVT (AVT40), cal. 7.62 x 54R, of Soviet manufacture.
Russian (U.S.S.R.), Stechkin, automatic pistol, model APS, cal. 9 x 18 (Makarov), of Soviet manufacture.
Sedgley, Glove Pistol .38 centerfire, S/N IRS-3580, single shot, marked MK2S, mfd. by Sedgley, Philadelphia, for the U.S. Marines.
Simonov-PTRS 1941, 14.5mm anti-tank rifle.
Smith & Wesson, Model 76 submachineguns, 9mm parabellum caliber, all S/Ns, originally manufactured by Smith & Wesson prior to 1975.
Smith & Wesson, Model 76 submachinegun, cal. 9mm parabellum, S/N range from U100 to U6100.
Solothurn Maschinengewehr, Model 38 (MG38), 7.92mm machinegun, mfd. between 1930 - 1935, by Waffenfabrik Solothur AG, Solothurn, Switzerland.
Solothurn, models S18-100, S18-1000 and S18-1100 20mm semiautomatic, antitank rifle.
Springer, .32 cal., Knife-Pistols, marked "C-16" and "C-12" on the left and right extractors, respectively, with "BAZAAR" marked on the left side of the blade. Some devices are marked "Solingen" on the left side of the blade and "Made in Germany" on the other.
* Springfield, .50 caliber M3 aircraft basic machinegun, S/N 1254946.
SS1-Sidewinder Machinegun, S/N Ex-016.
STAR, models AD, BD, MD, and PD, Selective Fire Pistols, any cal., produced prior to 1953.
Stevens, Offhand Number 35, .410 gauge shot pistols.
Stevens, Auto Shot Number 35, .410 gauge shot pistols.
Stoner, model 63 and 63A, machineguns, all variations, as originally manufactured from 1963 through 1970.
Suomi, model 26 and 31 submachinegun.
Swedish, model 37 and 39 submachinegun.
Swedish, Model 45, and Model 45B 9mm parabellum cal., produced in Sweden.
Swedish, model M-42, 20mm recoilless anti-tank rifles.
Swedish, 3.7mm Bofors anti-tank guns produced prior to 1946, including the following variations and designations: SwedishBofors, 37mm PvKan 34, Polish 37mm Armata przeciwpancerna wz. 36 (German captured designation 3.7 cm PAK 36(p), Danish 37mm antitank gun, designation unknown German captured designation 3.7 cm PAK 157(d), and British Ordnance QF 37mm Mk.1.
Swedish, model 21, 6.5mm, auto rifles mfd. prior to 1946.
Swedish, model 37, 6.5mm, auto rifles mfd. prior to 1946.
Swiss (Societe Industrielle Suisse) NKPO carbines, 7.65mm, with barrels less than 16".
Thompson, model 1927, semiautomatic carbine, caliber .45 ACP.
Thompson, submachineguns including all models, prototypes, & variations actually mfd. in or before 1945.
Thompson submachinegun, Model M1, .45 cal., mfd. by Auto Ordnance Corporation, West Hurley, New York, between 1985 and 1986, having S/Ns from M100A to M717A.
Thompson, model 1928, Korean War Commemorative submachineguns, cal. .45 produced by Auto Ordnance, West Hurley, NY.
Tippman Arms Company, .22 cal. machinegun set, consisting of a Model 1919-A4, S/N 333, Model 1917, S/N 333-A, Model .50 HB, S/N 333-B.
Tippman Arms Company ½ scale Browning Model 1917, water cooled, cal. .22 machineguns.
Tippmann ½ scale, Browning Model 1919 and .50HB machineguns, caliber .22.
Trejo, selective fire machineguns, caliber .22, mfd. in Mexico, circa 1952 to 1972.
Umbrella Gun, 6-shot pepper box revolver with dagger attachment extending from muzzle, combined with handle and umbrella, S/N 16, cal. .22 rimfire, with markings Brevette PT, mfd. by Marquis de Fabrique, France.
United Defense Supply Corporation, model U.D. 42, submachinegun, cal. 9mm parabellum, as mfd. by Marlin
Firearms Company, New Haven, CT.
* United States Field Gun, Model 1902, 3-inch caliber, produced prior to 1916.

SECTION IV

U.S. Arms Company, knife pistol, cal. .22, mfd. prior to 1934.
U.S. Browning machineguns (original), cal. .30, M1917 and M1917A1, mfd. by Colt's Patent Firearms Company, New England Westinghouse Company, Remington Arms, Union Metallic Cartridge Company, Rock Island Arsenal, High Standard Company, Savage Arms Corporation, Buffalo Arms Corporation, Frigidaire, AC Spark Plug, Brown-LipeChappin, Saginaw Steering Gear Division, and Kelsey Hayes Wheel Company.
U.S. Browning 1918 and 1918A2 BAR, mfd. by Colt, Winchester, Marlin Rockwell, New England Small Arms, IBM, and Royal Typewriter.
U.S. Browning model 1919, cal. .30, machineguns, produced by various U.S. Government contractors prior to 1946, all variations.
U.S. Browning model 1919A4 air cooled machinegun, produced by various U.S. Government contractors prior to 1946, cal. .30.
U.S. Carbine, cal. 30, M2, original U.S. Military issue, mfd. prior to 1946, and marked "M-2" on the receiver.
U.S. Johnson models 1941 and 1944 light machineguns, cal. .30.
U.S. M2 Browning, .30 cal. machineguns (ANM2), as mfd. by various U.S. Government contractors prior to and during World War II.
U.S. M 4 Survival Rifle, cal. .22 Hornet, as issued by the government, mfd. in the United States and Canada by Harrington & Richardson Arms Company.
U.S. Machinegun, Model T161, caliber 7.62, S/Ns EXP21, EXP25, EXP26, EXP27, EXP30, EXP32, EXP34, EXP35, EXP37, EXP38, EXP42, EXP45, EXP47, EXP47, EXP51, EXP53, EXP56, EXP57, EXP60.
U.S. Maxim machinegun, model 1904, caliber .30.
U.S. Model M3 and M3A1 submachinegun, cal. .45, original Guide Lamp or Ithaca Gun Co. manufacture only, mfd. prior to 1954.
U.S. M3 submachinegun, 9mm parabellum, as originally produced for the U.S. Government, having no manufacturer's name or S/N and originally marked U.S. 9mm SMG on the left side of the magazine housing, original production only.
U.S. Model M6, 37mm, gun, produced prior to 1946.
U.S. Model T-48 assault rifles, caliber 7.62mm, mfd. by H&R.
U.S. Models Ml, M9 and M18, cal. 2.36" Rocket launcher (Bazooka) and variations thereof.
U.S. Models 1915 and 1918 Vickers machineguns mfd. by Colt in cals. 30/06 or 11mm.
U.S. Mortar (original), cal. 60mm, M1 and M2, with mount M2.
U.S. M6 Survival rifle/shotgun, cal. .22/410 gauge, having 14-1/2" barrels, produced by various U.S. Government contractors.
* U.S. Pack Howitzer, Model M1-A1, 75mm caliber, S/N 336, manufactured in 1941.
U.S. Small Arms Company knife pistol, cal. .22 original manufacture produced prior to 1934.
U.S. 3" Anti-tank Gun, model M5.
U.S. 37mm Antitank guns M3 and M3A1, produced by various U.S. Government contractors prior to 1946.
U.S. 76mm Tank Cannon, Models M1, M1A1, M1A2.Vickers commercial water cooled machinegun, any configuration, any cal., British manufacture.
Vickers original military machineguns, all variations, all cals., mfd. prior to 1946.
Winchester model 92 Trapper carbines having barrel lengths of less than 16" (original Winchester manufacture only).

Frequently Asked Questions

Licensed Collectors

What modifications can be made on C&R firearms without changing their C&R classification?

The definition for curio or relic ("C & R") firearms found in 27 C.F.R. § 478.11 does not specifically state that a firearm must be in its original condition to be classified as a C&R firearm. However, ATF Ruling 85-10, which discusses the importation of military C&R firearms, notes that they must be in original configuration and adds that a *receiver* is not a C&R item. Combining this ruling and the definition of C&R firearms, the Firearms Technology Branch (FTB) has concluded that a firearm must be in its original condition to be considered a C&R weapon.

It is also the opinion of FTB, however, that a minor change such as the addition of scope mounts, non-original sights, or sling swivels would not remove a firearm from its original condition. Moreover, we have determined that replacing particular firearms parts with new parts that are made to the original design would also be acceptable—for example, replacing a cracked M1 Grand stock with a new wooden stock of the same design, but replacing the original firearm stock with a plastic stock would change its classification as a C&R item.

What is a firearm curio or relic?

Firearm curios or relics include firearms which have special value to collectors because they possess some qualities not ordinarily associated with firearms intended for sporting use or as offensive or defensive weapons. To be recognized as curios or relics, firearms must fall within one of the following categories:

1) Have been manufactured at least 50 years prior to the current date, but not including replicas thereof; or
2) Be certified by the curator of a municipal, State, or Federal museum which exhibits firearms to be curios or relics of museum interest; or
3) Derive a substantial part of their monetary value from the fact that they are novel, rare, bizarre, or from the fact of their association with some historical figure, period, or event.

[27 CFR 478.11]

Is there a specific license that permits a collector to acquire firearms in interstate commerce?

Yes. The person may obtain a collector's license; however, this license applies only to transactions in curio or relic firearms. The principal advantage of a collector's license is that a licensed collector can acquire curios or relics in interstate commerce.

[27 CFR 478.41(c), (d), 478.50(b) and 478.93]

How does one get a collector's license?

Submit ATF Form 7CR, Application for License (Collector of Curios or Relics), with the appropriate fee in accordance with the instructions on the form. These forms may be obtained from the Firearms and Explosives Licensing Center in Atlanta, Georgia, your local ATF office, or downloaded from ATF's Internet site (www.atf.gov).

[27 CFR 478.41(c)]

Does a collector's license afford any privileges to the licensee with respect to acquiring or disposing of firearms other than curios or relics in interstate or foreign commerce?

No. A licensed collector has the same status under the Gun Control Act (GCA) as a nonlicensee except for transactions in curio or relic firearms.

[27 CFR 478.93]

Does the Federal firearms law require licensed collectors to comply with State laws and local published

ordinances which are relevant to the enforcement of the GCA?

Yes. It is unlawful for any licensed collector to sell or deliver any firearm or ammunition to any person if the person's purchase or possession would be in violation of any State law or local published ordinance applicable at the place of sale or delivery.

[18 U.S.C. 922(b)(2), 27 C.F.R. 478.99(b)(2)]

Does a license as a collector of curio or relic firearms authorize the collector to engage in the business of dealing in curios or relics?

No. A collector's license only enables the collector to transport, ship, receive, and acquire curios and relics in interstate or foreign commerce, and to make disposition of curios and relics in interstate or foreign commerce, to any other licensee, for the period stated on the license. A collector's license does not authorize the collector to engage in a business required to be licensed under the Act. Therefore, if the acquisitions and dispositions of curios and relics by a collector brings the collector within the definition of a manufacturer, importer, or dealer, he shall qualify as such. A dealer's license must be obtained to engage in the business of dealing in any firearms, including curios or relics.

[18 U.S.C. 922(a) and 923(a)(1), 27 CFR 478.41(c)(d)]

What does "engaged in the business" mean?

The term "engaged in the business," as applicable to a firearms dealer, is defined as a person who devotes time, attention, and labor to dealing in firearms as a regular course of trade or business with the principal objective of livelihood and profit through the repetitive purchase and resale of firearms, but such term shall not include a person who makes occasional sales, exchanges, or purchases of firearms for the enhancement of a personal collection or for a hobby, or who sells all or part of his personal collection of firearms.

[27 CFR 478.11]

Are licensed collectors transfers of curio or relic firearms subject to the Brady law, including the provision for making background checks on transferees?

No. Transfers of curio or relic firearms by licensed collectors are not subject to the requirements of the Brady law. It is, however, unlawful to transfer a firearm to any person knowing or having reasonable cause to believe that such person is a felon or is within any other category of persons prohibited from receiving or possessing firearms.

[18 U.S.C. 922 (d), 27 CFR 478.32 (d), 18 U.S.C. 922(t)]

Is the transfer of a firearm by a licensed dealer to a licensed collector subject to the Brady law?

The Brady law does not apply to the transfer of a curio or relic firearm to a licensed collector, but a licensed collector who acquires a firearm other than a curio or relic from a licensee would be treated like a nonlicensee, and the transfer would be subject to Brady requirements.

Are licensed collectors required to execute ATF Form 4473 for transactions in curio or relic firearms?

No. Licensed collectors are only required to keep a "bound book" record.

[27 CFR 478.125(f)]

Are licensed collectors required to comply with the requirements that written notification be given to handgun transferees and signs be posted on juvenile handgun possession?

The requirement that written notification concerning juvenile handgun possession be given by licensees to a nonlicensee to whom a handgun is delivered applies to curio or relic handguns transferred by licensed collectors. Licensed collectors may lawfully dispose of curio or relic handguns away from their licensed premises. Therefore, the signposting requirement does not apply to licensed collectors since, in the case of collectors, a requirement to post signs at the licensed premises would serve no purpose because the premises is not a business premises open to the public.

[18 U.S.C. 922 (x), 27 CFR 478.103]

Can a licensed collector sell a curio or relic shotgun or rifle to a nonlicensed resident of another State?

Yes. A licensed collector is specifically authorized to sell a curio or relic shotgun or rifle to a nonlicensed resident of another State so long as 1) The purchaser meets with the licensee in person at the licensee's premises to accomplish the transfer, sale, and delivery of the rifle or shotgun; and 2) The sale, delivery, and receipt of the rifle or shotgun fully comply with the legal conditions of sale in both such States.

[27 CFR 478.96 (c)(1)]

Are licensed collectors required to turn in their acquisition/disposition records to ATF if their collector's license is not renewed or they discontinue their collecting activity?

No. The GCA requires the delivery of required records to the Government within 30 days after a firearm "business" is discontinued. A license as a collector of curios or relics does not authorize any business with respect to firearms. This is in contrast to firearm importers, manufacturers, and dealers who are licensed to engage in a firearms business. Therefore, the records required to be kept by licensed collectors under the law and regulations are not business records and are not required to be turned in to ATF when collectors' licenses are not renewed or collecting activity under such licenses is discontinued.

[18 U.S.C. 923(g)(4), 27 CFR 478.127]

May a licensed collector obtain NFA firearms in interstate commerce?

Only if the firearms are classified as curios or relics, are registered in the National Firearms Registration and Transfer Record, and are transferred in accordance with the provisions of the NFA.

[27 CFR 479.83 – 479.86]

May semiautomatic assault weapons, which have been classified as curios or relics, be imported?

Not unless they are being imported for sale to a government agency or law enforcement officer employed by such agency for official use. Because ATF will not approve an importation that would place the importer in violation of the law, ATF would not authorize the importation of semiautomatic assault weapons, even if classified as curios or relics, unless the importer provided evidence that the weapons were being imported for sale to a governmental entity or other exempt purchaser.

[18 U.S.C. 922(v)]

Made in United States
Troutdale, OR
11/03/2026

56847397R00086